Lean Manufacturing

Edited by Karmen Pažek

Published in London, United Kingdom

IntechOpen

Supporting open minds since 2005

Lean Manufacturing
http://dx.doi.org/10.5772/intechopen.92922
Edited by Karmen Pažek

Contributors
Ssafra Imen, Kaouther Ghachem, Shyam Sunder Sharma, Rahul Khatri, Venkataramanaiah Saddikuti, Saketh Saddikuti Venkat, Ganesh Babu Shanmugam, Sevilay Demirkesen, Kassu Jilcha Sileyew, Selamawit Gebreyohanis, Yash Dave, Stella Despoudi, Lokpriya Mohanrao Gaikwad, Vivek K. Sunnapwar, Vasileios Ismyrlis, J.P. Rishi, Sajit Jacob, Krishnamurthy Kothandaraman, Ruhet Genç, Luis Alfredo Avila-Lopez, María Marcela Solís-Quinteros, Carolina Zayas-Márquez, Teresa Carrillo-Gutiérrez, Bruno Vallespir, Anne Zougar-Amrani

First published in London, United Kingdom, 2021 by IntechOpen
IntechOpen is the global imprint of INTECHOPEN LIMITED, registered in England and Wales, registration number: 11086078, 5 Princes Gate Court, London, SW7 2QJ, United Kingdom
Printed in Croatia

British Library Cataloguing-in-Publication Data
A catalogue record for this book is available from the British Library

Additional hard and PDF copies can be obtained from orders@intechopen.com

Lean Manufacturing
Edited by Karmen Pažek
p. cm.
Print ISBN 978-1-83969-149-2
Online ISBN 978-1-83969-150-8
eBook (PDF) ISBN 978-1-83969-151-5

For EU product safety concerns:
IN TECH d.o.o., Prolaz Marije Krucifikse Kozulić 3, 51000 Rijeka, Croatia, info@intechopen.com or visit our website at intechopen.com.

Meet the editor

Karmen Pažek was born in 1976. She graduated from the Faculty of Agriculture, University of Maribor, Slovenia, in 2000. In 2001 she was employed at the same faculty as an assistant for the field of grassland management. Between 2000 and 2003, she enrolled in the master's study program in Agriculture Economics at the same faculty, and in 2003 she received her master's degree. In the same year, she enrolled in a doctoral study in Agriculture Economics at the Faculty of Agriculture and obtained the status of a research assistant. In 2006 she successfully completed her Ph.D. in Agriculture Economics. Since 2006 she has been habilitated at the University of Maribor, and Life Sciences (she has been a full professor since 2016) for the field of Farm management. She holds several courses at all levels of study. She is currently the head of the 1st-degree study Agriculture Economics and Rural Development and the Vice Dean for Education. Her research includes the development of decision support tools and systems for farm management (simulation modeling, multicriteria decision analysis, option models, risk management), the economics of agricultural production, and other modern methods of operational research.

Contents

Preface

Time is the most important element in any production system. As such, optimizing time is integral in industry and can be accomplished via the implementation of lean manufacturing. Enterprises that employ strategies and practices to increase total production efficiency, reduce production costs, and minimize waste are considered to be "lean." This book addresses important topics in the context of lean manufacturing that are crucial for the successful functionality of the production system.

This book is divided into two sections. The first section, "Lean Manufacturing Concepts in Enterprises," includes chapters focused on enterprise production with an emphasis on lean manufacturing. Chapter 1, "Application of Lean in a Small and Medium Enterprise", analyzes the direct effect of lean manufacturing practices on operational performance in the manufacturing industry. Chapter 2, "Lean and Kaizen: The Past and the Future of the Methodologies", discusses the adoption of lean manufacturing through effective lean practices depending on interpretations of past experiences and present information. Chapter 3 "Introduction to Lean Waste and Lean Tools", introduces the challenge of capturing, organizing, and disseminating throughout the aggregate business unit is a huge responsibility of upper management. Generally, in industry, some tangible and intangible factors exist in the form of non-value-adding activities that hinder the implementation of lean manufacturing processes. These are known as lean manufacturing barriers (LMBs). Chapter 4 "Effect of Lean Practices on Organizational Performance", focuses on new lean manufacturing approaches which, combined with management tools lead to more flexible and agile production and distribution processes in the textile industry. Because the life cycle in the textile and apparel industry is short, a new integrated approach to production and distribution planning is needed. In Chapter 5, "Enhancement of Textile Supply Chain Performance through Optimal Capacity Planning", the authors present linear programming techniques and integrating subcontracting activities as a good example for solving this problem. Poor productivity and efficiency in production are major problems for most industries relying on a heavy workforce. To prevent inefficient work practices, the construction industry has shifted its focus from the traditional approach to a more innovative one called lean construction. Chapter 6, "From Lean Manufacturing to Lean Construction: How Principles, Tools, and Techniques Evolved", discusses how lean construction aims to maximize value while minimizing waste. There are other advantages too; it creates safer, smoother, and more efficient processes to eliminate waste. The stability of the environment is a major problem in a turbulent environment. This leads to relatively long projects that are not compatible with a highly variable economic environment. The objective of the evolution management presented in Chapter 7, "Model-Based Enterprise Continuous Improvement", is to use enterprise modelling and all the benefits it brings in a framework that allows for more continuous improvement than is generally observed. It is obvious that effective utilization of resources is necessary, particularly when it comes to the manufacturing industry. Single Minute Exchange of Dies (SMED) is one of the classical methods used to reduce setup time. Chapter 8, "Single Minute Exchange of Dies: Classical Tool of Lean Manufacturing", discusses SMED applications in the gear industry. Chapter 9, "Lean Manufacturing as a Strategy for Continuous Improvement in Organizations", examines the changes faced in the Fourth Industrial Revolution that have generated challenges that will only encounter solutions through innovative efforts and industrial improvements as well as a radical change in our way of interacting. In the current revolution, there are digital changes that cause ruptures in social, economic, and political aspects, and the administrative process is part of this. Chapter 10, "Development of Integrated Lean Six Sigma-Baldrige Framework for Manufacturing Waste Minimization: A Case of NAS Foods Plc", examines how lean manufacturing increases quality and productivity in companies by reducing waste and therefore production costs; adapting favorably to the different innovation systems that are required today. Lean manufacturing has gained popularity in the past decade within the quality management domain. Chapter 11, "Analysis, an Anathema: Is That a Fervent Diatribe of Lean?", examines the fundamentals of sustainable lean manufacturing through Quality 4.0 with an understanding that manufacturers need it to build and sustain competitive advantage. Chapter x presents and discusses lean applications in the real world.

The second section, "Lean Manufacturing Practices and Environment," includes chapters connected with the environment and food production. Small and medium enterprises (SMEs) are facing the burden of non-equilibrium of the supply-demand chain along with global climate change. Several SMEs are looking for a substitute that can create a balance between performance and the environment. Despite numerous studies related to green and lean manufacturing, none clearly define the spheres of green and lean. In Chapter 12, "Lean Manufacturing towards Green Manufacturing Practices and Its Implementation in SME's", the authors explore the advancement of lean and green manufacturing and its impact on other sectors. The chapter also highlights the methodology adopted in implementing the same. Given growing ecological consciousness, environmental achievements of lean manufacturing also incorporate a strong economic relevance. Chapter 13, "Lean Manufacturing Practices and Environmental Performance", investigates the impact of lean manufacturing practices on environmental performance and the existing coherences between lean and ecologically oriented variables such as resource usage, energy consumption, and air pollution. The food sector has been criticized for its lack of sustainability and circularity due to the high levels of food and packaging waste as well as increasing costs. Although food supply chain entities have started to implement a circular economy and lean practices, the current efforts do not seem to be sufficient to achieve a circular and lean food system. Chapter 14, "Circular and Lean Food Supply Chains", discusses the possibility of creating circular and lean food supply chains.

Karmen Pažek
Chair of Agricultural Economics and Rural Development,
Faculty of Agriculture and Life Sciences,
University of Maribor,
Maribor, Slovenia

Section 1

Lean Manufacturing Concepts in Enterprises

Chapter 1

Application of Lean in a Small and Medium Enterprise

Venkataramanaiah Saddikuti, Saketh Saddikuti Venkat and Ganesh Babu Shanmugam

Abstract

Application of lean principles in manufacturing as well as services has been revolutionizing the operations for more than five decades. Many large as well as small enterprises have implemented lean and reported benefits in both direct and indirect activities of business. Due to advent of digital technologies and better understanding of process improvement approaches made lean much more effective across many sectors. In this chapter, we highlight various elements of lean and its application to a small enterprise in food processing sector in India. We draw some useful insights based on the implementation of lean and challenges faced by SMEs.

Keywords: lean, Toyota Production System, customer value, value chain, food processing, SMEs, Total Quality Management

1. Introduction

> *"One of the most noteworthy accomplishments in keeping the price of Ford products low is the gradual shortening of the production cycle. The longer an article is in the process of manufacture and the more it is moved about, the greater is its ultimate cost" [1].*
>
> *Henry Ford 1926*

Lean was identified at Toyota Production System (TPS) to eliminate or reduce waste or non-value added activities in the manufacturing system. It is also believed that application of Lean was implemented by Henry Ford at Ford Motors in 1920s. Lean is defined as by the National Institute of Standards and Technology Manufacturing (NISTM) Extension Partnership's Lean Network [1]: "A systematic approach to identifying and eliminating waste through continuous improvement, flowing the product at the pull of the customer in pursuit of perfection." It is a systematic approach for reducing different types of wastes which are constituting around 95% of the total waste. Important elements of TPS is given in **Figure 1**. TPS/Lean system was built on two major aspects (i) Elimination of Waste and (ii) Respect for people. TPS has three pillars (JIT, Continuous improvement and Jidoka) with fundamental blocks of standardized and stable process and level production. Some of the wastes in manufacturing environment are given in **Figure 2**. **Table 1** gives the brief description of eight wastes in TPS [2, 3].

The main objective of this chapter is to highlight various elements TPS and wastes in a typical manufacturing organisation and demonstration of application of

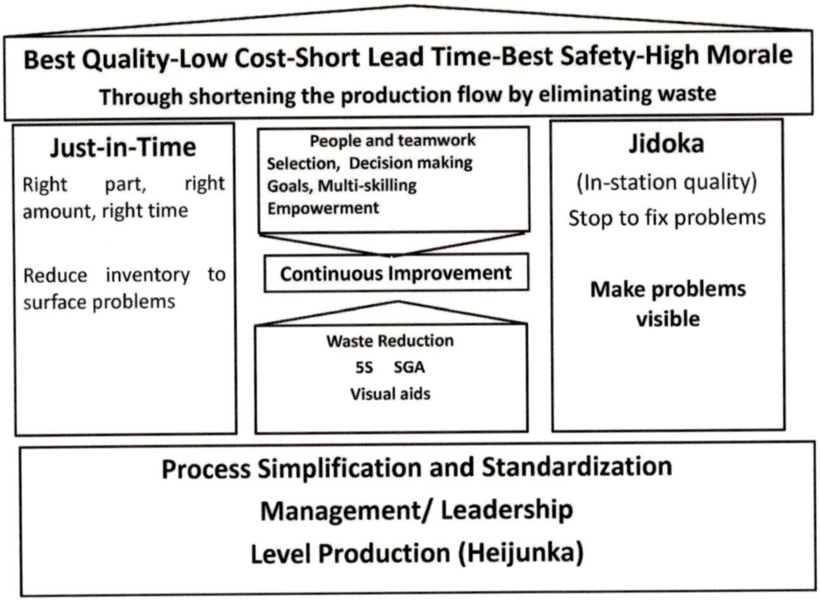

Figure 1.
Elements of Toyota Production System.

Figure 2.
Different types of wastes in manufacturing.

Sl No	Waste Type	Brief description and Examples
1	Overproduction	Producing more than required. It includes safety stocks, work-in-process, Finished goods etc
2	Waiting	Waiting for authorization, information, tools and equipment or operator. Example, waiting for machine, operator or handling equipment before or after completion of an operation at particular stage of the process.
3	Transportation	Distance travelled by various resources including parts, operators, handling equipment from point of availability to point of use. Example, an item travelling from storage to assembly or from vendor to manufacturer.
4	Over Processing	Processing more than required. Example, finishing more than specifications
5	Inventory	Any resource kept in reserve for future use due to uncertainties. This would include raw materials, WIP or finished goods.
6	Defects/Errors	Not meeting the customer specifications. Under size or over size dimensions of an item.
7	Excess Motion	Unnecessary motion or movement of products/resources due to poor layout or lack of coordination.
8	Underutilized People	Underutilization of highly skilled manpower. Example, high skilled person doing low skill job.

Table 1.
Different types of wastes in manufacturing.

lean at a small food processing enterprise. This chapter is organised in five sections. Section two highlights the methodology adopted in the chapter, section three describes building blocks and benefits of lean, section four application of lean at a small food processing organisation and finally conclusions in section five.

2. Methodology

In this study, we have adopted a generic method for literature search and industry practices in the area of lean manufacturing. Literature search has been carried out using key words like Lean, SMEs, Food processing, Toyota Production System, customer value, TQM from the databases like ABI, EBSCO, Google Scholar, etc. The search does not include other databases. Apart from these we have also used the case of an SME in food processing sector. The author's own research and consulting experience in the area of lean and transformation of SMEs in manufacturing and services.

3. Building blocks and benefits of lean

Lean/TPS uses various tools or building blocks which are proven and easy to implement in practice for minimizing or elimination of waste. Major building blocks of Lean include the following:

- Pull System
- Kanban

- Work Cells
- Total Quality Management
- Total Productive Maintenance
- Point-Of-Use-Storage
- Quick Changeover/SMED
- Batch Size Reduction
- 5S
- Visual Controls
- Concurrent Engineering

These are not only used in manufacturing organizations but also apply equally to service organisations. Summary of these are given in **Table 2.**

Many organisations around the world from manufacturing and service organisations have reported both operational and administrative benefits of successful implementation of Lean principles. According to NISM survey of firms implemented lean have reported operational and administrative benefits and are given below.

Operational benefits:

- 90% reduction in Lead Time (Cycle Time)
- 80% reduction in Work-In-Process Inventory
- 80% improvement in quality
- 75% improvement in space utilization
- 50% increase in productivity

Administrative Improvements includes the following:

- Reduction in order processing errors
- Streamlining of customer service functions
- Reduction of paperwork
- Reduction in staff required
- Improvement due to outsourcing of non-critical functions
- Reduction employee attrition
- Improved job standards

Sl No	Name	Brief Description
1	Pull System	Manufacturing based on customer demand/order, Make-to-order (MTO), in terms of exact requirement and time. This is against Push system. Capacity driven or Make-to-Stock (MTO).
2	Kanban	Producing based on accurate and timely information given in the form of a card where material is moved based on the timely information between stages. It can use single card Kanban or dual card Kanban system. Size and number of Kanbans are decided based on the volume of demand and production economics.
3	Work Cells	Lean or TPS focus on improved flow of material, manpower and other resources. The resources are organised around the requirements of the job done. Generally U-shaped layouts are more productive. It also helps better coordination and communication
4	Total Quality Management	Lean is built based on TQM fundamentals where every aspect of the organisation is very important. It recognizes the strength of the human resource and team work.
5	Total Productive Maintenance	It is based on proactive or preventive maintenance approach using knowledge and cooperation of people, vendors and other resources. It tries to identify and eliminate the breakdowns and improve the reliability of the system for improved throughput.
6	Point-Of-Use-Storage	This aspect of lean focus on keeping the required resources near the place of use. For example, the tools and equipment required at work center kept at the work center itself. Dedication of resources.
7	Quick Changeover/ SMED	It focuses on reduction of long changeovers which will be costly in terms of time and cost. It allows more frequent changes and smaller lot sizes.
8	Batch Size Reduction	Traditionally, manufacturing organisations used to manufacture to reduce the cost of set-up. In pull system, small batch size is more appropriate which will result in low WIP, low quality cost etc. Small batch size increased inventory turnover and better visibility, improved cash flow.
9	5S	5S is a systematic way of organizing the workplace.
10	Visual Controls	TPS heavily uses visual controls in almost every aspect of the business since it enhances the productivity, visibility and easy to understand at the executional level.
11	Concurrent Engineering	This approach helps in reducing the lead time drastically and utilizes cross-functional teams. This aspect particular helpful in reducing the time-to-market of new products/services. Some of the empirical results shows that around 50% decrease in the time-to-market.

Table 2.
Building blocks of Lean/TPS.

4. Application of lean at a small food processing organisation

This section highlights the implementation of Lean at a small food processing organisation in Southern part of India. The organisation was established in the year 2002 and gained a significant market share. The founder of the organisation is a first generation entrepreneur with high levels of enthusiasm and energy along with dedicated team of around 50 members including 10 members representing planning and execution [4].

4.1 The context

The organisation is a leading manufacturer and exporter of high-quality aquarium and pet food.

The organization was unable to increase the sales turnover and also struggling to maintain healthy profit margin despite that the organization catered to both Indian and export market. The founder of the organisation sought the help of one of the authors of this chapter for improving the market share and revenue. A five-member team consisting of internal members and external Lean transformation expert identified the following challenges.

- Sales Turnover is stagnant for the last three years
- Declining profitability
- Manufacturing cost is on an increasing trend
- No clarity on the losses in both manufacturing and sales function
- No management reviews and lack of data transparency among the team members
- No proper coordination among Team on deliverables
- Founder get involved in all day to day decision making

4.2 Solution methodology adopted

Internal team along with external Lean transformation expert identified various areas for improvement using brainstorming sessions and listed the following areas for improvement.

- Training of the core team on Quality Management and Lean project
- Awareness workshop for the senior executives
- Development of overall system with suitable metrics and fixing of responsibilities for each functional area of the business.
- Identified the waste in plant utilization,
- Efficiency, and quality issues
- Implemented Quick changeover techniques to reduce changeover losses between food products
- Identified the constraints that affecting the equipment efficiency
- Set the key performance metrics (KPI's) for customer deliveries, cost, and quality and reviewed every month along with Business Head
- Implemented sales and operational key performance reporting on a daily, weekly basis to avoid communication problems
- Analysed and implemented right inventory norms for raw material and finished goods to improve stock availability and reduce inventory

- Identified and implemented cost-saving projects in plant and administrative areas
- Identified the staffing levels and gaps and recruited suitable skilled personnel in the areas of shortage
- Improved the communication process among operations, sales, and finance team to take decisions on time
- Identified the utilization of floorspace in the plant and movement of the products inside and outside the organisation

4.3 Application of lean tools & process

The team has adopted and designed transformation process using well proven lean management tools and processes including the following:

- Performance Measurement and Management at business and operation level
- Application of Lean tools and techniques
- Designed suitable inventory management systems for raw material and finished goods
- System for Performance reporting on daily, weekly and monthly basis
- System for communication at all levels of the organisation
- Metrics for supply chain management at internal and external to the organisation.

4.4 Business result achieved

- 40% reduction in changeover times
- 40% reduction in inventory value without affecting deliveries
- 33% Reduction in number of shifts (from 3 shifts to 2 shifts)
- 30% Improvement in plant capacity utilization;
- 30% increase in production tonnage with reduced working hours
- 15% increase in sales turnover and increase in delivery performance by around 10%
- 15% (appr) increase in operating profit (from negative to positive)
- 5% reduction in overall manufacturing cost

4.5 Recommendation for sustainability and future growth

Based on the detailed study and implementation, the team has suggested the following for future growth.

- Training and development of workforce in all functional areas with right caliber of professionals
- Monitoring and evaluation system for financial performance and cash flows
- Expanding into new markets
- Structured review mechanism on business and functional performance on regular basis
- Accountability improvement through organizational structure/role clarity and management practices
- Streamlining New Product development process w.r.t new markets and sales plan

5. Insights and conclusions

Implementation of Lean has benefited both small and large business organisations in manufacturing and services. Lean systems/practices are built on the strong empirical evidence of both operational and administrative benefits. Lean focuses on elimination of waste by utilizing the dynamics of team work and strength of people and processes. Implementation of lean has proved in improving quality of products and services across all industries and organizational boundaries and revolutionizing the operations [5]. There are many other studies like [6–10] focused on various aspects of lean implementation. However, the implementation of lean requires certain level of certainty/stability in the system, discipline and culture at inter and intra organizational level. Further, studies can focus on the impact of socio, economic and technological determinants of lean in different sectors of business.

Author details

Venkataramanaiah Saddikuti[1*], Saketh Saddikuti Venkat[2] and Ganesh Babu Shanmugam[3]

1 Indian Institute of Management Lucknow, India

2 Amazon Development Center, Chennai, India

3 Winning Minds Solutions, India

*Address all correspondence to: svenkat@iiml.ac.in

References

[1] National Institute of Standards and Technology Manufacturing (NISTM) Extension Partnership's Lean Network https://www.creconline.org/blog/clients/national-institute-of-standards-technology-manufacturing-extension-partnership/ [Accessed on 2021-02-20]

[2] Wilson, L. (2010). *How to Implement Lean Manufacturing?* New York: The McGraw-Hill Companies, Inc.

[3] Liker, Jeffrey K. (2004). The Toyota Way: 14 Management Principles from the World's Greatest Manufacturer. McGraw-Hill.

[4] Ganesh Babu S and Venkataramanaiah Saddikuti (2019), Implementation of Lean at a food processing unit, IIM Lucknow, India.

[5] Duncan E and Ritter R (2014), Next Frontiers for Lean, McKinsey Quarterly, February 2014.

[6] Eswaramoorthi, M., Kathiresan G. R., Prasad P. S. S. and Mohanram P V (2011), A survey on lean practices in Indian machine tool industries, International Journal of Advanced Manufacturing Technology, Vol.52, pp:1091-1101.

[7] Panizzoloa R, Garengoa, P., Sharma M K and Gorec, A (2012), Lean manufacturing in developing countries: evidence from Indian SMEs, Production Planning & Control, Vol. 23, (10-11), pp: 769-788.

[8] Chiarini, A (2011), Japanese total quality control, TQM, Deming's system of profound knowledge, BPR, Lean and Six Sigma Comparison and discussion, International Journal of Lean Six Sigma, Vol. 2(4), pp: 332-355.

[9] Ma Ga (Mark) Yang M G., Hong, P and Modi, S B (2011), Impact of lean manufacturing and environmental management on business performance: An empirical study of manufacturing firms, International Journal of. Production Economics, Vol. 129, pp: 251-261.

[10] Dora, M, Kumar, M and Gellynck, X (2016), Determinants and barriers to lean implementation in food-processing SMEs – a multiple case analysis, Production Planning & Control, Vol. 27 (1), pp:1-23.

Chapter 2

Lean and Kaizen: The Past and the Future of the Methodologies

Vasileios Ismyrlis

Abstract

Lean and Kaizen improvement methodologies have been in the entrepreneurship spotlight for a long time. They can be adopted by any kind of enterprise, and they succeed in producing better long-term results, improving their performance, but most important, influencing the philosophy of the organizations implemented. In this research, many case studies and success stories of companies implementing Kaizen or/and Lean methodologies, or even the new Lean Kaizen methodology, will be introduced. We attempt to evaluate the performance of Lean and Kaizen implemented companies and distinguish the elements that made the difference. Maybe, it is some specific tool, or an aspect in the culture that was enhanced, since the implementation of these business process improvement methodologies. Finally, thoughts and estimations will be presented, regarding the future of these methodologies, in the unstable and rapidly changing economic environment.

Keywords: Lean, Kaizen, Lean Kaizen, performance, process improvement

1. Introduction

This chapter presents the methodologies of Lean and Kaizen, their contribution in the enterprises' field and their future potential contribution to the field of management.

1.1 Lean (production, manufacturing or management)

First of all, lean manufacturing, will be referred, as it seems to be a methodology, that was keenly embraced and wields a great acceptance in the management (theoretical and practical) world. It started with the efforts of Toyota automobile company, while some believe that Ford motor company, was also a great influence in creating its concept.

Its main scope and its achievement is the elimination of waste and this can lead in an increase in productivity.

There are many proven cases of continuous improvement with the implementation of lean thinking methods and tools. However, it is not a concept that impose specific rules or tools to be implemented, but rather it is a philosophy that encourages efforts in order to achieve its main goal, which is to eliminate waste.

1.2 Kaizen

Kaizen is more of a philosophy than specific technique or methodology, yet it has also affected and changed seriously the minds of the managers. It utilizes many analytical techniques to succeed in its mission, as value-stream mapping and the5 why's.

It has been introduced that these two methodologies have provided valuable assets in the field of management. It is notable that in the last years, Lean is also referred as Lean Management, representing the value that the methodology has produced in the management field. Moreover, and even though these two above methodologies, were introduced in the manufacturing field, they have managed to expand in all business industries and it seems that have achieved much in improving organizational performance. They are highly appreciated in the theoretical and practical field of management. They are considered (many times along with Six Sigma), as business process improvement methodologies, since they aim at improving all the processes in an organization. They perform activities in order to gather data, to track and evaluate all the functioning processes, and of course, they apply a continuous improvement effort.

However, in the last years, many other aspects have emerged in the economic status, like the constant economic uncertainty, which seems to be critical and threatens even the existence of many companies. Hence, the answer to be answered is if and with which manner could these methodologies keep up their efforts and produce sustainable solutions for the enterprises.

To answer these questions many writings from experts in the field were looked into the literature and their views were registered. In the conclusion, the future aspects of these methodologies will also be presented.

2. Introduction to Kaizen and Lean

2.1 Lean

Lean is called by many names as lean manufacturing, lean production, or lean thinking. New terms are the lean management and lean industry.

The main focus of the Lean methodology, is to eliminate waste in order to obtain more resources dedicated in finding ways to satisfy the customers. It intends to intervene to the value stream of the organization, in order to improve or eliminate every unneeded process that waste resources. It is also said that lean production was founded on the idea of Kaizen.

It has a great history and it is always linked to automotive industry. Many ideas that form the lean philosophy, were created by car companies like Ford and Toyota. Sometimes lean thinking is referred as a synonymous to Toyota or Toyota production system.

Author [1] has presented the Toyota Production System, which has been defined as a method which focuses on defining and eliminating non-value added activities or waste in all systems and processes [2]. One of its approaches was the Just-in-time (JIT) methodology, which acquired the necessary resources when they were exactly needed and helped in solving many material flow problems.

Authors [3], in their book 'Lean thinking: Banish waste and create wealth in your corporation', that introduced lean to a broader audience, have managed to extend the concept in a general perspective. The same authors, define lean as: "a way to specify value, line up value creating actions in the best sequence, conduct these activities without interruption whenever someone requests them, and perform them more and more effectively".

Lean is a multi-faceted concept which was identified and coined to explain the success of the "Japanese Way of Working" that enhanced their increased competitiveness at the time [4].

Components of the "Lean Idea" include:

- operations concepts, such as zero inventories [5], Just-in-Time (JIT) [6] and small lot sizes [7];
- the underpinning of robust quality procedures exemplified by Total Quality Management (TQM) and Total Productive Maintenance (TPM); and,
- a method of working that encourages empowered employee participation which challenges the over-bureaucratic top-down, function orientated organizational structures that had traditionally dominated many "Western" organizations [8].

This view of Lean was endorsed [9], categorizing the components of Lean into four "bundles":

- JIT bundle
- TQM bundle
- TPM bundle and
- Human Resource Management (HRM) bundle.

To be successful in implementing all these Lean facets in a coordinated, coherent manner, strong leadership and a clear alignment with organizational strategy over many years is required.

The 4P's model of lean are:

In the book of [10], the 14 principles for continuous improvement are categorized in four pillars (P's):

- Philosophy
- Process
- People
- Problem solving.

The above framework is created and implemented in Toyota company.

2.1.1 Customer value

Lean emphasizes in the provision of value to the customer and there are three types of value:

- Values added: contribute directly to the needs of customer.
- Non-value add: no contribution to the needs of customer.
- Non-value add: (but necessary or essential non-value add)

2.1.2 The eight wastes

According to lean philosophy, the non-needed (not-adding value) activities are considered a waste. These are the following:

- Defects
- Overproduction
- Transport (moving of products)
- Waiting
- Excessive Inventory
- Motion (moving of people)
- Processing (unwanted process steps)
- Skills (lacking)

The main wastes were seven. Skills is an new addition.

Lean is much depended on the tools and methodologies that it utilizes. Some lean strategies that seem more successive for the concept and can be implemented in many different ways (e.g. merged), are [11]:

- 5S
- Automation
- Continuous flow
- Continuous improvement
- Kan-Ban
- Kaizen
- Six Sigma
- Total Quality Management (TQM)
- Value stream mapping (VSM)
- Work standardization
- Zero defects concepts
- Lean thinking
- Work in progress
- Flexible manufacturing system

2.1.3 Benefits from lean adoption

The benefits from the implementation of lean can include many aspects, like the minimization or even elimination of waste, less work load, qualified and skilled workers, zero delays, saving time, reduction of costs, etc. [12].

Lean manufacturing changed the way that industry worked in the era of mass production and it presents many differences from the traditional manufacturing way. The major differences include [11]:

Higher flexibility, higher customer satisfaction, higher empowerment, shorter lead time, the inspection is performed in a 100% level and by workers, the inventories are produced per demand, the batch size is small and continuous, pull scheduling is implemented.

However, lean has not only domained the manufacturing field; it has also managed to enter in many more sectors, like service, trade etc. It has been also accepted as a new management system and a new term 'lean management', was created [13]. Hence, lean seems that can play a vital role in the management theory and practice, in order to produce a sustainable future for the enterprises.

2.1.4 Barriers and disadvantages from lean

Lean implementation or the attempts to implement it, presents some drawbacks like the below:

- Every new concept and change in the workplace it is not easy to be accepted by the human workforce.
- Lean is not considered easy to be implemented in practice [14].
- The main scope of lean is to maintain industry stable without any disorders, but this is difficult to be applicable.
- The barriers created from the new concept, prevent the workers to perform their duties normally. However, lean thinking can contribute in overcoming those barriers, by creating a cooperative environment [14].
- Not every industry seems ready to accept the drastic changes of lean in production and quality [15].

2.2 Kaizen

Kaizen is a Japan-oriented strategy (also referred as culture, philosophy, approach, or methodology), which literally means continuous improvement (CI). It manages to involve all the workforce of an organization in its activities (e.g. Kaizen events, suggestion system) and it also highlights the importance of the workplace as the center of all actions, activities and processes. One of its main advantages is that it does not induce financial burdens to the organizations.

Its main philosophy is to produce small changes, which when taken together they can have a large impact. It utilizes the continuous improvement approach in every aspect of the organization.

It aims to involve workers from multiple functions and levels in the organization in working together to address a problem or improve a process. It requires skilled and well trained workers to achieve its scope.

It was first captured and implemented in big manufacturing sites, like the Toyota motor company. However, its philosophy was suggested that it can be implemented in every human life activity [16, 17]. Anywise, its main idea is that everyone and everything can change to the better, doing small steps.

It promotes process-oriented thinking [18] and continuous improvement of the standard way of work [19]. It is an endless effort for improvement, involving everyone in the organization [20]. The actions of mobilizing staff and encourage them to participate, could make them also responsible and able to contribute to the company's development [21].

Sometimes it is identified as a management approach (like TQM, Lean manufacturing, or company wide quality control), while other scholars considered it as a group of techniques and tools for cutting waste and finally others, appraised highly its success to intensify staff participation through its suggestion schemes [22].

The scholar that managed to introduce it in a formal aspect, was Imai, with his two books [16, 23], although the method still lacks a detailed explanation that would clarify better its theoretical context [22, 23]. Kaizen forms an umbrella that covers many techniques including Kanban, Total productive maintenance (TPM), Six Sigma, Just-in-time (JIT), suggestion system etc. [16]. According to [16], Kaizen is a continuous improvement process involving everyone. Broadly defined, Kaizen is a strategy to include concepts, systems, and tools within the bigger picture of leadership involving and people culture, all driven by customer [23]. Its success in uncovering a problem, making it visible, looking for its root causes and then eliminating them, was of extreme importance in the development of the manufacturing sector in countries such as Japan and Korea [24].

2.2.1 Kaizen actions

One of the actions that Kaizen implements is the Kaizen event, which is a five-day (or six) team workshop defining specific goals for an area that requires improvement. A team leader will lead this event and will include training, data collection, brainstorming, and implementation. At the end of the event, the team leader will create a follow-up plan and a report to be submitted to management.

A typical Kaizen event may include the following

- Define goals and provide the necessary information
- Evaluate the current status and create a plan for improvements
- Implement planned improvements
- Review and fix what does not function
- Report results and determine any follow-up items.

The above cycle is also referred as PDCA (Plan, Do, Check and Act). This cycle is a vital part of the Kaizen philosophy.

2.2.2 The Key Players of the Kaizen Team

Kaizen requires the support of an appropriate team with the right Team Roles. There are the following team roles.

- Champion: The champion is the person driving the train. It is normally a senior manager or executive who can defy barriers and inspires.
- Facilitator: Part coach, part trainer, and part leader, this person is typically well-versed in Lean and brings substantial experience to the table. At any given time, a facilitator may be overseeing several kaizen events in different stages.
- Team Leader: The team leader is in charge of the event and does most of the planning and preparation under the supervision of the facilitator. The team leader is often the manager, supervisor, or engineer in charge of the process being improved.
- Team Members: Team members are picked from the work area, from the stakeholders, or from the company at large. The best teams combine a variety of experience and skills. Teams typically range from 5–10 people depending on the size of the project.
- Stakeholders: Kaizen events influence a lot of people. Those people are known as stakeholders and should be included in decisions about the project.
- Support Team: Kaizen events often require support that goes beyond what team members can perform. This often falls on the facilities team and on IT.

2.2.3 Kaizen umbrella (tools and techniques)

Kaizen philosophy needs assistance to achieve its scope and therefore it includes many weapons (they called the Kaizen umbrella), such as the following:

- Total quality control
- QC circles
- Suggestion system
- Automation
- Kanban
- Just-in-time
- Zero defects
- New product development
- Quality improvement
- Total productive maintenance
- Small-group activities

Of course many of the above Kaizen activities, could also be part of the lean methodology, which sometimes includes a Kaizen project.

2.3 Lean-Kaizen

Lean-Kaizen is a new suggested approach of quality improvement in the literature that combines the two methodologies, which as already has been refereed, many times are implemented together and it is not such a surprise to see them combined together. However, it is also introduced as a new approach and will be presented as such, in this chapter.

The Lean-Kaizen technique, as a novel one, is composed of two basic words i.e. Lean and Kaizen which implies continuous elimination of waste through small-small improvements [25]. It is adopted for waste identification and elimination; it helps industry to be lean [26, 27]. It is a systematic way that focuses on continuous improvement of the process, productivity, and quality of the product by suggesting effective and efficient Kaizen events [10]. Leanness can also be defined in terms of efficiency and effectiveness of the manufacturing system [28].

The adoption of the Lean-Kaizen approach improves the organization output by solving problems through identifying and implementing small improvements in process, product, and system [29, 30]. So, the Lean-Kaizen approach is required to be implemented in order to produce quality products by eliminating waste Muda) in the entire system of the organization [31].

Quality is understood as a measure of excellence or a synonym of zero defects, zero deficiencies or absence of variations in the product by many industries. In order to achieve the desired product quality, the quality system performance is continuously monitored and evaluated for the sake of constant improvements of customer satisfaction, morale and reliability [32].

3. The contribution of the methodologies

3.1 Lean

3.1.1 In which industries

Lean and its strategies can eliminate all types of industrial waste [33]. Lean manufacturing has as a goal to eliminate waste and it succeeds in it without having to define additional requirements of resources [34, 35]. However, it seems that except the manufacturing sector, which gave birth to the concept of lean, many more industries have been profited from its power.

3.1.2 In which fields

Lean manufacturing contributed in improving manufacturing operations, protecting the industrial jobs and lift customer satisfaction [36].

The appropriate implementation of lean, improves the quality and the productivity and reduces the amount of inventory and work processes [37].

Except the improvement in productivity, lean manages to level up customer and employers satisfaction [36, 38].

The study of [39], presented the implementation of lean in various types of industries and it managed to achieve various types of waste reduction, manufacturing system design parameters and business value achievements.

In their study [40] concluded that despite the resistance to change in public organizations, the implementation of lean succeeded in the optimization of resources and the simplification of processes.

In a study in health care services, [41], noticed that waste was eliminated and quality was maximized, benefiting the customers.

Public sector seemed to be a nice field to implement the thinking of lean, as there is many waste and the needs for better quality are more than before.

Therefore, continuous improvement approaches have been formally applied in the public sector all over the world, in an attempt to improve service quality and streamline processes, often in response to cuts in public expenditure budgets imposed by governments.

New Public Management (NPM), a new theory, emerged as the supporting doctrine to this policy, that advocated the imposition in the public sector of management techniques and practices drawn mainly from the private sector, as according to NPM greater market orientation would lead to better cost-efficiency, with public servants becoming responsive to customers, rather than clients and constituents, with the mechanisms for achieving policy objectives being market driven.

A new effort that contributes in examining lean in the public sector is the "Lean in public sector" project (http://leaninpublicsector.org/). Launched in 2007, the aim of LIPS was to extend lean project management to public sector construction projects. Its scope is to include the application of lean thinking to government operations generally so that new facilities support, new and more effective ways of delivering government services.

Some of the successes of this project, include:

- Introducing the lean management philosophy and methods to Australia's project alliancing.

- Following the pioneering work of the Finland's Transportation Agency, a range of Finnish government organizations has successfully applied lean and Integrated Project Delivery (IPD) principles to over 35 projects since 2009 with more on the way.

- At the 2013 conference it was announced that the European Commission ruled against a challenge to the contract award of one of those Finnish projects, thus providing proof that integrated project delivery is legal under EU construction procurement regulations.

- In the US, the University of California, San Francisco (UCSF) has led in the development and testing of alternative contract structures and methods of aligning commercial interests, and this without multi-party contracts, which are not currently allowed for the university system.

- The California state university system and many community college systems are also successfully applying lean concepts and methods within the limits of current regulations.

3.2 Kaizen

Some examples of Kaizen implementation and success are presented in the **Table 1**.

It can be concluded that Kaizen has also been implemented in organizations of all business industries and provided valuable solutions.

Authors	Industry	Tools, actions	Results
[42]	Manufacturing foods product	5s Technique, team training	Decrease in quality rejections, reduction in change over times and increase in manufacturing activities.
[43]	Industrian technology	Kaizen approach and lean thinking	Reduction in space used, material handling costs, lower scrap rates.
[44]	Manufacturing industry	Six sigma, JIT	Eliminate waste, increased sales.
[45]	Manufacturing	Kaizen event, inventory management Kaizen,	Reduced process time
[46]	Public agricultural organization	Kaizen project, 5S,	Process improvement, shorten work processes, decrease n financial expenses
[47]	Semiconductor industry	Kaizen technique	Cost reduction,
[48]	Automobile assembly production line	Set-by-step kaizen procedure	Elimination of major functional problem, reduction in quality rejections, elimination of rework processes.

Table 1.
Kaizen implementation examples and results.

4. Conclusion

Having realized the potential of Kaizen and Lean, even big consulting organizations have dedicated many writing in presenting and exploring the methodologies. Consulting companies are organizations that provide professional services in other companies in the fields of marketing, financing, ICT, logistics, business plans etc. Some of the biggest consulting companies worldwide are: Deloitte, McKinsey, EY, Boston consulting company.

For example Deloitte has attempted to connect Lean with Industry 4.0* and has presented many other cases of lean implementation in several fields. Some links of relevant articles follow:

https://www2.deloitte.com/us/en/insights/focus/industry-4-0/digital-lean-manufacturing.html

https://www2.deloitte.com/us/en/blog/human-capital-blog/2020/lean-strategic-planning-design-thinking-agile-what-does-it-all-mean-in-becoming-exponential-hr.html

https://www2.deloitte.com/content/dam/Deloitte/ca/Documents/life-sciences-health-care/ca-en-life-sciences-health-care-lean-in-health-care.pdf.

https://www2.deloitte.com/us/en/insights/focus/industry-4-0/digital-lean-manufacturing.html.

* Industry 4.0 is the concept of creating a digital enterprise by establishing digital technologies and integrates them with advanced production and operation techniques.

McKinsey, also a big consulting organization, has presented some analytical guides and presentation of the methodologies. Many of them are evident in financial institutions and the links of some follow:

https://www.mckinsey.com/business-functions/operations/our-insights/the-work-of-leaders-in-a-lean-management-enterprise

https://www.mckinsey.com/~/media/mckinsey/industries/consumer%20packaged%20goods/our%20insights/the%20consumer%20sector%20in%202030%20trends%20and%20questions%20to%20consider/2014_lean_management_enterprise_compendium.pdf

https://www.mckinsey.com/~/media/mckinsey/dotcom/client_service/financial%20services/latest%20thinking/reports/lean_management_new_frontiers_for_financial_institutions.pdf

https://www.mckinsey.com/business-functions/operations/our-insights/next-frontiers-for-lean

Digital lean

Digital lean is an example of the integration of digital technologies and lean principles It utilizes lean theory to decrease the waste in digital technology actions and processes.

Digital lean uses Industry 4.0 and other digital tools to create the appropriate information for all operations and processes. As data come in a high frequency way, it can be managed and directed in the appropriate resources.

Digital lean can be a valuable asset and some of its achievements are: reduced costs, improved quality and higher return on investment, compared with any other methodology that is implemented individually.

Lean Industry 4.0

A new concept deriving from the combination of lean and Industry 4.0 is presented.

The main scope of lean is to reduce waste in the value chain, focusing on client's value and strengthening the role of the employees in all this process [3].

On the other hand, the basis of Industry 4.0 is the ability to quickly collect, process, analyze and exchange large data sets between machines. Thanks to modern technologies such as: Cyber-Physical Systems (CPS) or Internet of Things (IoT), it is possible to react faster and more flexibly to existing problems, but also to more efficient value creation processes, while reducing costs [49].

A suggested solution is the combination of the above concepts, in order to solve the problems that modern production faces. Despite, the significant differences, between the two concepts, they seem to have the same goal, to increase added value [50].

4.1 The present and future of the methodologies

Lean can provide solution in many fields and it is evident that even service companies embrace and appreciate it [51]. Lean reached functions that previously seemed quite difficult to transform. (management principles once known as lean manufacturing [52].

The new concept of lean management, which introduces the values of lean in the modern management, has been adopted by many organizations. It provides a roadmap that holds the organization (and the workflow) stable and able to solve all of the derived problems. The primary winner of these efforts is the customer and then consequently the organization.

Lean seems not to be a static methodology. It is still developing and could be a valuable solution for many enterprises [53]. Lean can be fitted in the rapid changing world, which seems to be even more intense after the COVID-19 crisis. The Toyota lean business system has managed to deliver: better quality, productivity, customer focus, innovation, employee engagement, profitability and even environmental sustainability [54]. Organizations should concentrate on involving all employees in the continuous improvement organizing appropriately the value stream and offer the ideal products and services for the customers. Money and profit should not be their main incentives, as customers have improved demands and there is intense competition.

Questions that challenge current entrepreneurship could be, how well large, modern organizations work as almost as old as management [52]. Problems that could affect organizations are: slower growth, debt burdens, aging workforces, mismatches between worker skills and available jobs.

There are successful organizations that attain a state of continuous improvement. Their performance is consistent in the short and long run. However, it is not always effective for every organization to imitate best practices.

Hence, lean management system [13] could be in the forefront of the management field, in order to provide valuable solutions. Its main aspect, which is to improve material flows, could be a valuable asset.

Lean can contribute in the appropriate integration of the technology field in the organizations, as it can provide valuable information with the customers' feedback [54].

The lean system could assist information management system to solve their information flow problems. If the information provided is the essential and appropriate, the information system could be benefited and improve its performance.

The information gathered with many ways and directed appropriately, could be an instrument to link this information with direct customer needs and inform the relevant departments in an organization.

From all the cases and the thoughts about Lean reported in this study, it is evident, that it is a philosophy that can be an asset in the management field and the question is if the organizations could make the right choices and be benefit from its positive aspects.

Author details

Vasileios Ismyrlis
Hellenic Statistical Authority, Edessa, Greece

*Address all correspondence to: vasismir@gmail.com

References

[1] Ohno, T., 1988, Toyota production system: beyond large-scale production, CRC Press.

[2] Vinodh, S., Arvind, K. R., and Somanaathan, M., 2011, Tools and techniques for enabling sustainability through lean initiatives, Clean Techn Environ Policy, 13, pp. 469-479.

[3] Womack, J. P., Jones, D. T., & Roos, D. 1990. Machine that changed the world. Rawson.

[4] Hu, Q., Mason, R., Williams, S. & Found, P. 2015. Lean implementation within SMEs: a literature review. Journal of Manufacturing Technology Management, Vol. 26 Iss: 7, pp.980-1012, 26(7), 980-1012.

[5] Hall, R.W. 1983. Zero Inventories. New York: McGraw Hill.

[6] Assessing changes towards lean production, Christer Karlsson and Pär Åhlström, 1996, International Journal of Operations & Production Management, Vol. 16, No. 2 1996, pp. 24-4.

[7] P. BURCHER, S. DUPERNEX, R. GEOFFREY, "The road to lean repetitive batch manufacturing: modelling planning system performance", International Journal of Operations and Production Management, 16 (2) 1996, pp. 210-221

[8] Hines, P., Found, P., Griffiths, G. and Harrison, R. 2010. Staying Lean. New York:Productivity Press.

[9] Shah, R. and Ward, P.T. 2003. Lean manufacturing: context, practice bundles, and performance. Journal of Operations Management, 21(2), 129-149.

[10] J.K. Liker, The Toyota Way: 14 Management Principles from the World's Greatest Manufacturer Hardcover, 2003, McGraw Hill,

[11] Chahal V., & Narwal M.S, 2017, An empricial review of lean manufacturing and their strategies, Management Science Letters vol. 7, pp.321-336.

[12] Abdollahi, M., Arvan, M., & Razmi, J. 2015. An integrated approach for supplier portfolio selection: Lean or agile? Expert Systems with Applications, 42(1), 679-690.

[13] Emiliani, B. (2013), The Lean Professor: Become a Better Teacher Using Lean Principles and Practices, The CLBM, LLC, Wethersfield, CT.

[14] Roslin, E. N., Shamsuddin, A., & Dawal, S. Z. M. 2014. Discovering Barriers of Lean Manufacturing System Implementation in Malaysian Automotive Industry. In Advanced Materials Research (Vol. 845, pp. 687-691). Trans Tech Publications.

[15] Harris, G., Stone, K. B., Mayeshiba, T., Componation, P. J., & Farrington, P. A. 2014. Transitioning from teaching lean tools to teaching lean transformation. Journal of Enterprise Transformation, 4(3), 191-204.

[16] Imai, M. 1986. Kaizen-the key to Japan's competitive success. New York, NY: Random House.

[17] Wittenberg, G. 1994, "Kaizen, the many ways of getting better", Assembly Automation, Vol. 14 No. 4, pp. 12-17.

[18] Hammer M, Champy J and Tathan R L 1993, Reengineering the Corporation:A Manifesto for Business Revolution, Harper Collins, New York.

[19] Chen J C, Dugger J and Hammer B 2000, "A Kaizen Based Approach for Cellular Manufacturing Design: A Case Study", The Journal of Technology Studies, Vol. 27, No. 2, pp. 19-27.

[20] Malik S A and YeZhuang T 2006, "Execution of Continuous Improvement

Practices in Spanish and Pakistani Industry: A Comparative Analysis", IEEE International Conference on Management of Innovation and Technology, Vol. 2, pp. 761-765, Singapore.

[21] Bessant, J. 2003, High Involvement Innovation, Wiley, Chichester.

[22] Suárez-Barraza M.F. & Rodríguez-González, F.G. 2015 Bringing Kaizen to the classroom: lessons learned in an Operations Management course, Total Quality Management & Business Excellence, 26:9-10, 1002-1016, DOI:10.1080/14783363.2015.1068594.

[23] Imai, M. 1997. Gemba Kaizen. New York, N.Y.: McGraw-Hill.

[24] Singh, J. and Singh, H. 2009, "Kaizen philosophy: a review of literature", The IUP Journal of Operations Management, Vol. 8 No. 2, pp. 51-72.

[25] Suarez-Barraza, M. F., Smith, T., and Dahlgaard-Park, S. M., 2009, Lean-kaizen public service: an empirical approach in Spanish local governments, The TQM Journal, 21(2), pp. 143-167.

[26] Prashar, Anupama, 2014, Redesigning an assembly line through Lean-Kaizen: an Indian case, The TQM Journal, 26(5), pp. 475-498.

[27] Suarez-Barraza, M. F., Ramis-Pujol, J., and Kerbache, L., 2011, Thoughts on kaizen and its evolution: Three different perspectives and guiding principles, International Journal of Lean Six Sigma, 2, pp. 288-308.

[28] Singh, Bhim, Garg, S.K., and Sharma, S. K., 2010, Development of index for measuring leanness: study of an Indian auto component industry, Measuring Business Excellence, 14(2), pp. 46-53.

[29] Panwar, Avinash, Jain, Rakesh, and Rathore, A.P.S., 2015, Lean implementation in Indian process industries – some empirical evidence, Journal of Manufacturing Technology Management, 26(1), pp.131-160.

[30] Singh, B., Garg, S. K., and Sharma, S. K., 2010, Scope for lean implementation: a survey of 127 Indian industries, International Journal of Rapid Manufacturing, 1(3), pp. 323-333.

[31] Bhuiyan, N., and Baghel, A., 2005, An overview of continuous improvement: from the past to the present, Management Decision, 43(5), pp. 761-771.

[32] Sunil Kumar, Ashwani Kumar Dhingra, Bhim Singh, 2017, Implementation of the LEAN-KAIZEN approach in fastener industries using the data envelopment analysis, Mechanical Engineering Vol. 15, No 1, 2017, pp. 145-161

[33] Belekoukias, I., Garza-Reyes, J. A., & Kumar, V. 2014. The impact of lean methods and tools on the operational performance of manufacturing organisations. International Journal of Production Research, 7543(July 2014), 1-21. http://doi.org/10.1080/00207543.2014.903348

[34] Bhamu, J., & Sangwan, K. S. 2014. Lean manufacturing: literature review and research issues. International Journal of Operations & Production Management, 34(7), 876-940.

[35] Vamsi, N., Jasti, K., & Kodali, R. 2014. A literature review of empirical research methodology in lean manufacturing. International Journal of Operations & Production Management, 34(10), 1080-1122. http://doi.org/10.1108/IJOPM-04-2012-0169

[36] Singh, B., Garg, S. K., Sharma, S. K., & Grewal, C. 2010. Lean implementation and its benefits to production industry. International Journal of Lean Six Sigma, 1(2), 157-168.

[37] Gupta, S., & Jain, S. K. 2013. A literature review of lean manufacturing. International Journal of Management Science and Engineering Management, 8(4), 241-249.

[38] Rose, A. N. M., Md. Deros, B., & Ab. Rahman, M. N. 2013. A study on lean manufacturing implementation in Malaysian automotive component industry. International Journal of Automotive and Mechanical Engineering, 8(1), 1467-1476. http://doi.org/10.15282/ijame.8.2013.33.0121.

[39] Yudha Prasetyawan, Mokh Suef, Nauval Rifqy, Intan Oktasari Kusuma Wardani, 2019, Manufacturing strategy improvement based on lean methodology, IOP Conference Series Materials Science and Engineering 508(1):012095.

[40] Renato Lopes da Costa , Tiago Resende Leandro Ferreira Pereira, , Álvaro Lopes Dias, Jose Santos., 2020, Public Sector Shared Services and the Lean Methodology: Implications on Military Organizations, September 2020, Journal of Open Innovation Technology Market and Complexity 6(78):1-13.

[41] Rubin Cohen , 2018, Lean Methodology in Health Care, June 2018Chest 154(6), DOI: 10.1016/j.chest.2018.06.005.

[42] Lee M 2000, "Customer Service Excellence Through people motivation and Kaizen", IEE Seminar, "Kaizen: from Understanding to Action" (Ref. No. 2000/035), Vol. 5, pp. 1-21.

[43] Lee S S, Dugger J C and Chen J C 2000, "Kaizen: An Essential tool for Inclusion in Industrial Technology Curricula", Journal of Industrial Technology, Vol. 16, No. 1, pp. 1-7.

[44] Ashmore C. 2001, "Kaizen and the Art of Motorcycle Manufacture", Manufacturing Engineer, Vol. 80, No. 5, pp. 220-222.

[45] Palmer V S 2001, "Inventory Management Kaizen", Proceedings of 2nd International Workshop on Engineering Management for Applied Technology, pp. 55-56, Austin, USA.

[46] Dehghan R, Zohrab M, Momeni A and Hoseini M 2006, "Continuous Improvement Approach in the Agriculture Sector", Asian Pacific Productivity Conference, Seoul National University, South Korea, available on http://www.fdp.ir

[47] Kikuchi K, Kikuchi T and Takai T 2007, "Method of Overall Consumable Effectiveness", IEEE International Symposium on Semiconductor Manufacturing, pp. 1-4, Santa Clara, USA.

[48] Chandrasekaran M, Kannan S and Pandiaraj P 2008, "Quality Improvement in Automobile Assembly Production Line by Using Kaizen", Manufacturing Technology Today, Vol. 7, No. 3, pp. 33-38.

[49] Oztemel, E., & Gursev, S. 2018. Literature review of Industry 4.0 and related technologies. Journal of Intelligent Manufacturing, 31, 127-182. https://doi. org/10.1007/s10845-018-1433-8.

[50] Prinz, C., Kreggenfeld, N., & Kuhlenkotter, B. 2018. Lean meets Industrie 4.0 – A practical approach to interlink the method world and cyber-physical world. Procedia Manufacturing, 23, 21-26. https://doi.org/10. 1016/j.promfg.2018.03.155. Current trends and future perspectives, Cogent Business & Management, 7:1, 1781995.

[51] Lean management: new frontiers for financial institutions McKinsey, 2011, https://www.mckinsey.com/~/media/mckinsey/dotcom/client_service/financial%20services/latest%20thinking/reports/

lean_management_new_frontiers_for_
financial_institutions.pdf

[52] McKinsey: the lean management enterprise, 2014a. https://www.mckinsey.com/~/media/mckinsey/industries/consumer%20packaged%20goods/our%20insights/the%20consumer%20sector%20in%202030%20trends%20and%20questions%20to%20consider/2014_lean_management_enterprise_compendium.pdf

[53] McKinsey,2014b, article-Next frontiers for lean, https://www.mckinsey.com/business-functions/operations/our-insights/next-frontiers-for-lean

[54] D.T.Jones, 2020, https://planet-lean.com/lean-management-post-covid19/)

Chapter 3

Introduction to Lean Waste and Lean Tools

Shyam Sunder Sharma and Rahul Khatri

Abstract

In the turbulent and complex business environments, many Indian SMEs are facing stiff competition in the domestic as well as in the global market from their multinational counterpart. The concept of lean has gained prominence due to the fact that the resource based competitive advantages are no longer sufficient in this economy. Hence, lean is no longer merely an option but rather a core necessity for engineering industries situated in any part of the globe, if they have to compete successfully. Lean Manufacturing (LM) which provides new opportunities to create and retain greater value from the employee of the industry based on their core business competencies. The challenge of capturing, organizing, and disseminating throughout the aggregate business unit is a huge responsibility of the top management. The success of any industry depends on how well it can manage its resources and translate in to action. The adoption of lean manufacturing through effective lean practices depends on interpretations of past experiences and present information resides in the industry. Generally, in an industry, some tangible and intangible factors exist in the form of non-value adding activities which hinder the smooth lean implementation are known as lean manufacturing barriers (LMBs).

Keywords: Lean, waste, kaizen, manufacturing

1. Introduction

In the present worldwide situation, manufacturing industries are primarily handling difficulties from two directions. First, cutting edge manufacturing ways of thinking are arising, while the current techniques are getting outdated. Second, consumers demand is changing in very short of time. The clients have become more demanding for inventive product in short timeframe and at less cost. Basically, to adapt up to such difficulties, the idea behind manufacturing industries' these days is to capture the customer demand while limiting waste [1]. Subsequently, manufacturing firms working in such quick changing in customer demand in competitive market. In last thirty years, manufacturing industries are introducing lean thinking. The lean manufacturing word means to minimize the industrial waste or to eliminate the waste and improve the benefits of manufacturing industries by smooth production flow [2]. Without lean practices any industry cannot be successful in the present-day situation due to globalize competition in market with low cost, high quality, and shorter delivery time. It is very difficult for engineering industries to shift from traditional system to lean system. a troublesome undertaking to move from a conventional assembling framework to a lean assembling one. This change makes attention both employee and method.

Lean manufacturing is a management strategy that tries to make industries more competitive, minimize the manufacturing cost by eliminating the industrial waste and increase the productivity of an organization [3].

1.1 Research background of lean

Afterward World War II Japanese industries were confronted with the difficulty of immense deficiencies of skilled manpower, money, and material. These difficulties that industries of japan were challenged with compared from those of their Western partners. This introduces "lean" thinking in manufacturing industries. Toyota Motor Company runs by Toyoda identified that American manufacturers were manufacturing better than Japanese manufacturing industries; during the 1940's American industries were beating Japanese manufacturing industries. To take an action for progress timely Japanese pioneers like Shigeo Shingo, Toyoda Kiichiro, and Taiichi Ohno, they are receiving the challenge to improve the production system by eliminating the industrial waste, they have developed a new manufacturing strategy that is called "Toyota Production System," or "Lean Manufacturing." Taiichi Ohno, accept the responsibility to improve efficiency at Toyota is the primary force to develop Toyota production system. Ohno drew upon certain thoughts from the Western countries, and especially from Henry Ford's book "Today and Tomorrow." Ford's stirring production lines for continuous material flow developed the basics for TPS. After research, the TPS was refined somewhere in 40's and 70's, is still emerging today all over the world. The crucial thought of this system is to maximize the resource utilization and minimize the inputs that cannot be enhanced any value to a product that is a waste.

To contend in the present furiously competitive market, United States manufacturing industries has understand that the mass production idea must be modified to the lean manufacturing. An assessment that was done by MIT of transformation from mass production toward lean manufacturing, as explained in the book "The Machine That Changed the World" [4] arise the US companies from their nap. The assessment highlighted the extraordinary achievement of Toyota and pulled out the enormous gap that developed between the Japanese and Western engineering firms. The thoughts came in the mind of US industrialists on the ground that the Japanese industries developed, manufactured, and delivered items within less manpower, less investment, less floor utilization, less time, instruments, raw material, and overall investment cost [5].

1.2 What is lean?

The fresh transformation in engineering products and service division has made extraordinary difficulties for US firms. The consumer focused and exceptionally aggressive market has delivered old-fashioned of management that was not enough to overcome these complexities. These factors present a key test to industries to seek for new methods to survive in competitive global market. While a few industries keep on developing based on financial steadiness, different firms fight because of their absence of understanding of the change in consumer mentalities and cost practices. To avoid the present situation and to turn out to be more valuable, many industries implemented lean principles in their organization and perform well in global market [6].

Waste exclusion, cost drop, and employee encouragement are the basic ideas behind the lean manufacturing system, which has been implemented in Japanese companies for many years. The Japanese thinking of making business is completely distinct from the thinking that has been dominant in United States for a long time.

The typical western belief was that the just way to get turn a profit to apply it to the cost of production to reach the preferred sales price. The Japanese method, on the contrary assumes that the generator of the sale price is client. The more consistency you build into the manufactured goods higher the cost that consumers pay. The distinction among the price of the goods and this price is what decides the profit [7, 8]. To minimize cost, raise investment, get in more revenues, and remain competitive in a rising international market, the lean manufacturing discipline is to function in all parts of the value stream by reducing waste. The value stream is explained as the specialized activities needed to plan, order and supply a specific product or value within a supply chain [9, 10]. As of Womack describe it the term "lean" indicates a system that utilizes less with respect to output, to produce the equivalent outputs as those generated be a conventional mass manufacturing system, which adding more varieties to the final consumer [11]. This theory of business goes by various names. Agile production, just-in-time-production, synchronous production, world-class production, and continuous flow are all concepts that are used in contrast with lean production. The resounding theory of lean manufacturing, therefore, is to minimize costs by continuous improvement, which would ultimately reduce the cost of services and goods, thereby increasing profits.

"Lean" focuses on the removal or reduction of waste ("muda", the Japanese word for waste) [12, 13] and on optimizing or allowing maximum use of activities that add value from the perspective of the consumer. Quality is equal to something that the consumer is willing to pay for in a product or service that follows, from the viewpoint of the customer. The reduction of waste is also the central concept of lean manufacturing.

1.2.1 The 8 wastes of lean

The aim of lean is to abolish the waste from the production process. It is very important to identify the eight waste before digging it. Waste is in the least action or activity that will not enhance any cost to the product, or we can say, waste is any unwanted process that will reduce the value of the product and customer do not want to pay for that. Taiichi Ohno identified the initial seven types of waste that was called Muda in japan [12]. Transportation, inventory, movement, waiting, overproduction, overprocessing and defects are seven types of waste identified by Taiichi Ohno. The acronym 'TIMWOOD' also applies to them. The eighth waste was invented by western industries in 1990s, and that was unused of workers talent or 'Skill' of workers was later added. Therefore,' TIMWOODS' [14] is generally referred to as the 8 wastes [15].

1. **Transport:** Unwanted movement of the product during manufacturing. It is caused due to unplanned layout and product are unnecessary move from one workstation to other. In addition, excessive movement causes fatigue, wear and tear of product and equipment's [16–18].

2. **Inventory:** Over production or semi-finished product to convert into finished product. Sometime customer is not receiving the order or customer is canceled the order. So, this type of products is store and called waste. The advantage of inventory is that some time vendor will offer discount on large amount of purchasing. For maintain large inventory manpower and store cost is also involved and there is chance of product damage. Over procurement, work in progress (WIP) or the production of excessive goods than the customer demands may trigger surplus inventory. Certain inventory countermeasures take in procuring raw materials only when appropriate amount needed, reducing buffers between production stages, and establishing a queue system to avoid overproduction [15, 17, 18].

3. **Motion:** Workers are moving from one workstation to the other workstation without necessity and the manufacturing lead time is increase. This type of unwanted motion is considered as waste. Any excessive movement of workers, vehicles, or machinery requires waste in motion. Running, raising, reaching, bending, stretching, and shifting are part of this. To improve the working conditions for workers and improve health and safety standards, repetitive motion activities should be eliminated. Some motion countermeasures consist to make sure that the tools material is place near machinery in well organized manner [19].

4. **Waiting:** These are time delay and idle times during which value is not added to the product. If the machines, men, and material wait it is waste of these resources and it demoralizes the employees. The waste of waiting includes: 1) Operator is waiting for his turn and not receive material on time. 2) Machines are idle due to line unbalance [19, 20].

5. **Overproduction:** Excess of production over consumption. In market demand is less compare to the consumption, but industries are manufacture more to reduce the manufacturing cost. In this case inventory cost is increase and money is also block. So, it is considered as a waste. Overproduction means manufacturing additional goods via a 'push production mechanism'. Three countermeasures to develop overproduction. Firstly, by use of 'Takt Time' confirms that the production rate among workstations is continue. Secondly, reducing idle time like loading and unloading, setup times. Thirdly, reduce the WIP by using a pull or 'Kanban' system [19, 21, 22].

6. **Over-processing:** Over-processing will increase machining time, material handling time and add more process steps. Due to over processing the cost of the product is increased that will pay by the customer. For reducing over processing on products, consider standard job specifications for manufacturing. Prior to starting work, always think to the customer and produce product quantity as per the requirements of the customers and try to reduce the unnecessary operations and manufactured quantities where it is required [19, 22].

7. **Defects:** The product is not manufactured as per the specifications and tolerances given by the customer. Those products are rejected in quality inspection and consider as waste. Product/material will reject when the product/material is not suitable for use. Due to defective product/material it will loss of money and defective piece will not be reused [23–25].

8. Skills - The 8th Waste: This waste was not developed by Toyota, this 8th waste - the waste of human skills - is well known to many individuals. Also explain as no utilization of manpower skills, creativity, efforts consider in the 8th waste. This waste is developed when management not identify the skills of his workers in the organization. Employees is just following the boss order and do work as per the boss instructions. It is very difficult to optimize the process without taking help of frontline workers. This is because the worker who perform the job on shop floor is recognize the problems first and he has the solutions for that problem [14].

1.2.2 Identifying and eliminating the 8 wastes

Perceiving that they exist and giving a proficient system to characterizing them is the initial step to slashing waste. Value Stream Mapping (VSM) is a tool of Lean Management to assess the current state and to design a likely state. This outlines the progression of information and substance as they emerge. VSM is an effective strategy to plan the process involved, outwardly show the connection manufacturing process and to recognize nonvalue added and value-added activities. Utilize the VSM to characterize waste and proceed in view of the end client. Work in reverse to the beginning of the production process from the end client. Record cases of the eight waste in the process and construct a methodology to eliminate or limit them. Keep on provoking the staff to discover more waste and reliably build up their strategies. Draw in with and bring out their thoughts for change from the forefront staff. They will grow more trust in their critical thinking abilities as the group keeps on limiting efficiencies and waste decrease turns out to be important for their regular everyday practice after some time.

1.3 The 9 principles of lean manufacturing

Assembly work is categorized by short development cycles and batch sizes continuously decrease, Although the number of categories of goods and the models are still growing. Constant pressure to cut manufacturing lead times precedes to these needs and really makes the mix difficult, also for the highly imaginative producers. The capability to react quickly needs evolving buyer requirements usage of production systems that it is possible to re-configure and extend the fly that can fit, and advances in methods for assembly without having any initial output obsolete investment [26].

Lean production, An Approach That depends heavily on versatility and flexibility. Organization of the office is an exceptional Starting point for businesses who want to take a new look at their present Methods for production. Lean approaches are also worthy of study, since big capital is removed by them dedicated equipment outlays until automation becomes completely, needed. The idea of lean manufacturing, indeed, represents a big departure from such a famous automated factory the past few years. The "less is better" Manufacturing policy leads to a widely condensed, strikingly uncluttered, environment which is carefully calibrated to the environment manufacturer's specifications. Goods are generated in response, one at a time, to the specifications of the customer rather than of the batch produced for inventory. The target is to only generate the amount used and no more [27].

The number of parts is produced, it can change procedures, it is appropriate to handle various components and allow full utilization of workers, services, and floor area. The intrinsic versatility in manual assembly therefore, cells are superior to automated ones. This maximum requirement flexibility makes distinctive requirement on the lean work cell and the elements compose a lean work cell. Admittedly, the lean solution is not the only the solution to all production issues. But it does deliver a versatility that is special solution for more complex assembly commodities. This guide explains 9 essential descriptions Lean principles of development that should be assist you in evaluating lean manufacturing solutions for your own.

1. Continuous Flow:

   The lean work cell's chosen to form U-shaped workcell. In order of method, each subprocess is linked to the next. And an employee within the U, minimum movement to move is needed the workpiece or one-piece assembly toward the next workstation. Ultimately, one of the targets of the slim workcell is to remove all movement with non-value-added; hence its U-shape. Where, when the procedure has been completed by the employee, he it just turns around and is back on the move. The workpiece may be carried from one piece to another. Operation with value applied to the next one. There are times, however when the workpiece or the fixture which holds the workpiece is too heavy and between workstations must be manually moved [28]. While it is possible to transport very heavy components on belt conveyors, manual push conveyors of gravity or gravity are suitable for moving the components between workstations. Theirs' The minimal complexity makes it easy for them to support and reduces time. Moreover, they are easy to attach to end-to-end, making it quick to switch inside a workcell workstations. The bent U-shaped "corners" a working cell can pose a problem. As they may serve as a possible dead space, they may act as a mini storage room, thereby facilitating a storage area going back to batch manufacturing. Alternatively, the use of a ball roller transfer should encourage the movement of parts through the corners and the U-shape [29, 30].

2. Lean Machines/Simplicity:

One-at-a-time from continuous-flow another aim of lean manufacturing is it is necessary to produce each one, the workstation is designed to match a nominal covering. The Minimum the envelope guarantees the removal of excess of flat space at the workstation or workstation that machine [28]. This is done to prevent the risk of components or subassemblies being stored from the computer. Components stock increases "work in method and outcomes in " batch processing, which then defeats the goal of lean. In addition, smaller workplaces and devices of minimal size remove unnecessary steps taken by the worker between Via subprocesses [31].

Ultimately, valuable floor space can be saved by sizing workstations correctly machines and the implementation of uniform machine bases or workstations for all processes should be avoided, while tempting for the sake of conformity and standardization. Every base machine to optimize assembly subprocesses, which in most cases may differ from workstation to workstation, the workstation or workstation should be built. For virtually every structural material, this customisation can be accomplished. However, to save on costs and to minimize the environmental issues associated with the disposal of inflexible welded steel structures, material that is reconfigurable and reusable should be given priority. The modular characteristics of extruded aluminum and bolt-together systems make them suitable for lean manufacturing principles to be applied. In addition, constant enhancement as a method, all workstations and work cells need to be simple to alter [32].

3. Workplace Organization:

The desired outcome of a smooth, uninterrupted flow of finished workpieces is a lean workcell, correctly planned. Nothing here this flow can be slowed or stopped quicker than the tool failure or misplacement. Thus, all, applications used on a workplace must have a holder on their own. Exactly, there are as many tool holders as there are tools, so that the deficiency of a tool is quick observed [33]. Using an integrated tool holder device for each instrument with a particular holder that is ideal. If it is possible to add holders quickly, to a workstation or taken away from it, this it adds to the workstation's versatility and enhances its usefulness in a lean production technique. Backup tools, to reduce downtime, at any automated workstations, they should also be available. These instruments should be equipped to being out of the way of the worker before a failure this happens at an automated workstation. In the maximum advantage is tool holding frameworks that allow instruments to swing or slide [30].

4. Parts Presentation:

Naturally, the workcell will require additional components during the average work shift. In a lean workcell, traditional techniques of resupplying workstations are not useful. With the minimum number of interruptions, each worker can go about his job. Each part should also be delivered from outside the work cell to each workstation. The use of gravity feed conveyers or bins suits the lean work-cell's streamlined nature. Parts bins should be filled from at the back (outside the work area of the work cell) so that production can be continued without interruption by the worker. Gravity transports the components to the area of reach of the worker. Bins can be reconfigurable as well. The containers using a key stud in the picture to secure them in place [30]. When reconfiguring the workspace, bins are conveniently stackable and provide the ultimate in flexibility.

While bins are suitable for small parts, larger parts are needed for many assemblies. In bins or boxes, these can be shipped. Again, without entering the work-space, the components should be sent to the workcell. This function is served well by gravity feed conveyors. An additional gravity feed conveyor can be placed in the reverse direction if scrap, or containers must be removed from the cell. Lift assist devices are recommended in instances where pieces are very large. With mechanical, pneumatic, or hydraulic control, heavy parts or boxes of parts can be loaded onto a case lifter and lifted to the correct working height.

5. Reconfigurability:

A lean workcell that is properly built must be easy to reconfigure. In fact, it is a must to be able to adjust the process and go from good piece to good piece as fast as possible. Faster the changeover, less time is lost in production. Switching can be done in a matter of seconds with a strong quick-change fixture. As the situation requires, a variety of different fixtures may be kept at the workstation and swapped. At times, a lean cell must be rapidly attributed to process shifts or other variables to accommodate assembly of a new product, reconfigured or even relocated. In the ability to transfer each part of the work cell rapidly becomes extremely essential if a computer or workstation needs to be changed. The versatility required for rapid and efficient changeover is given by lockable casters on machines or workstations [28, 34].

6. Quality:

A reduction in quality concerns is one of the consequences of one-at-a-time production. Visual inspection by the worker will check that it is correctly assembled when each component is made. They should be installed on the computer or workstation if verification is necessary via gages. And they can be replaced quickly. Fast release of fixtures is a must using star knobs or locking levers. There will be a time when it is not easy to address a quality issue. A defective method or a malfunctioning computer could be the root of the problems with consistency. In the case of a defective process, the structural framing scheme allows for improvements in a minimum amount of time, no matter how big. Once again, in limited time, bolt-together construction addresses a big issue [34]. A malfunctioning machine can also be easily replaced, particularly if it is fast. When the lean cell is constructed, disconnections for all pneumatic or electric lines are given. Furthermore, in the lean cell, there should be no pneumatic or electrical contacts between machines. These would slow the machines inside the cell from changing. If the system has been removed from all power sources, if installed on lockable casters, it can be transported easily. Ease of reconfiguration and swapping eliminate any inability on the part of the employee or management to attempt to “Make do” with “almost” accurate devices or processes. This adjustment in Attitude can contribute greatly to the development of true quality [35].

7. Maintainability:

A further requirement of a lean cell is ease of operation. In a pull-through system, long down periods cannot be tolerated. The product must be generated while consumer demand exists. The ultimate in keepability is given by a modular structural framework. Components may be removed in a matter of seconds [34]. The design of bolt together ensures that computer stands, part presentation equipment or workplaces can be repaired in seconds. In a limited amount of time, even whole computer bases can be restored. Also, the systemic framing scheme provides for all machine bases, guards, a source for common components, workstations and with standardized elements, maintaining a structure requires a minimum number of resources. Three or four basic hand tools are necessary to construct or restore any structure with a structural framing system [36].

8. Ease of Access:

All required work elements can be installed in easily available locations using an aluminum mounting system as the basis for a lean cell, since each side is a possible mounting side. For productive work, parts bins, instruments, shelves, and fixtures may all be placed in the ideal spot. The T-slot on the surface of the framing device often enables if clearance space is critical, swift repositioning of pneumatic or hydraulic parts. Components can be rapidly attached to any workstation and quickly repositioned to ensure each worker's usability [34]. Additionally, with simple hand tools, whole guards or individual panels can be removed easily, allowing service technicians to conduct maintenance in a matter of minutes [37].

9. Ergonomics:

The worker must, eventually, be shielded from ergonomic issues. Each lean work cell properly designed must be ergonomically designed. It is always necessary to maintain work at the ergonomically correct height in the work cell. While it is sometimes not considered, a design for the average height of the worker is also a requirement. Since average heights vary from country to country, it is important to easily change the height of a computer or workstation if there is a risk of a it is possible to ship workstations from country to country [34, 38].

1.4 Lean tools and techniques

Several industries introduce Lean by seeing Lean as a series of 'tool'. For a while, this could be helpful, but in the long run, it will not be enough. Behavior is developed by defining values like as dragging the Andon chord when a difficulty arises, but it continuously does this, always expects it, and always supports it [39]. Lean techniques are the base of lean thinking and the most common applied techniques are listed below:

1. 5S

It is the most common methods used in lean management. Starting the Lean journey with 5S, however, might not be a good idea. Although 5S is simple to incorporate, it has improve the efficiency and quality, it can also be a distraction from real goals or simply clean-up [39]. A 5S program's real goals should be:

- To lower waste
- To enhance variant
- To increase productivity

It is necessary for senior management to be supportive before introducing the 5S Lean technique.

Since 5S may be ideal model for many organizations, even if they understand the concept of it. Using the model methodology is one way of helping workers grasp 5S. It targets small section of the shop floor and implements 5S there. Before any consideration is given to moving to another location, the 5S should be identify every detail. The primary reason for doing this is to inspire employees to look and assess the outcomes of 5S with their previous way of working. Since 5S would be the better option compare to the older ways, the workers would be ready and able to proceed to other regions and eventually the whole business with it.

i. **Sort:** Everything sorted in the work area. First, they categorized what is required for manufacturing and what are not required. Those that are not needed in the work area or serve no purpose must be discarded immediately. The company can choose to red tag products when in doubt. Red tag is a sticker indicate the date of object and then it is discarded if the object is not used up to the date [17].

   The products are sorted accordance with the use. High use equipment's are kept close (perhaps daily) as possible to staff so that they do not waste time reaching them. Those that are used less often are positioned slightly further (perhaps once a week) so it is easily reachable to the workers, but not very near as to compete with the use of regularly used objects. Lastly, those that are seldom used are kept furthest away (per-haps once a month).

   The sorting should be carried out regularly, maybe single time in a month but it is habit [18, 40].

ii. **Set in order:** The set deals with each item's location. Each item should be place in the manner that it can be easily available for everyone and everyone knows that where it is placed. Two methods used to identify the product for all employees where color coding and labeling on the product. Whenever there are some products, parts or instruments shift, this stage should be repeated [17, 18, 40].

iii. **Shine:** Everyday, the work area should be kept physically clean, workers also checked the working area that everything is placed in proper manner and if it is outplaced then it can be fix instantly. One technique is used to clean shop floor in every five-minute routine basis on each day (this process should be standardized for getting best results). The cleaning and tidying equipment's are properly arranged and regularly maintained. 'Cleaning is testing' implies the incorporation of both. You are not just washing up, you are looking for any abnormality's and their root causes [17, 18, 40].

iv. **Standardized:** For the first 3Ss, expectations must be established to confirm that the employees do what the business requires from them. "Standard work aims to create repeatable, reliable and capable processes and procedures". The greatest norm is one that employees consider to be so strong and consistent that the workers are followed the given process plan and they do not divert in some other way (or do the process in some other way) [17, 18, 40]. For the introduction of the 5S to be a success, these standards need to be well managed.

v. **Sustain:** All staff should make a habit of the first four Ss and must also continually strive to use and improve them. Audits are supported and enhance the values of 5S to uplift [17, 18, 40].

2. Just-in-Time (JIT)

JIT is a lean technique based on waste reduction and productivity growth. Waste can be defined as any action which does not add any value to the manufactured goods. Excess lead times, overproduction, and scrap are common examples of waste [17]. Instead of moving goods based on expected demand, JIT can be considered as a 'pull' operation based on client demand [40]. JIT's primary aim is to "produce and transport what is needed, when it is needed, amount needed, in the shortest possible lead time" [41]. "In summary, JIT is based on the concept of supplying raw materials just when required and producing products just when required".

3. Kaizen

The most well-known Lean approach is Kaizen. The combination of kai and zen, meaning "change" and "good" is Kaizen. This is what we have simply translated as "continuous improvement". For Kaizen implementation no initial cost is required or with in very less money it can be give big profits. Neither it cannot change the floor layout, nor it is need any advanced technology [17, 18, 40, 42].

4. Kanban

Kanban the Japanese word means "sign" or "card" This is the main technique used for continuous work flow between the work stations. It is used to identify the condition of product and what operations are carried out on the products and who is the operator. Kanban will maintain the flow of product from start to the end [40, 43].

5. Poka-yoke

"Poka-Yoke is fool-proofing technique for error prevention and elimination". This approach is not restricted to being used only in production but can also be used in office activities (such as post office, clinics etc.). Poka-yoke helps an industry to avoid the occurrence of a problem or flaw, or to interrupt a procedure immediately when a probation occurs. The clutch in a car is a normal and daily instance. The vehicle will not start until the clutch is pressed [44].

6. 5 whys

Sakichi Toyoda would have designed the Lean system of the "five whys". It is one of the significant approaches that Toyota uses to solve problems. The theory is to evaluate the problem before the root cause or causes are found, not to stop at the first cause of a problem (the first why). In fact, it is more of a theory than a cause analysis tool since it is not sufficiently organized nor 'accurate' (why 5 and not 4, 6? In the 2nd, the root cause can be quite well discovered) [44].

7. Andon:

The Japanese origin term is the mixture of the two symbols 行 (go) and 灯 (light) that can be translated as "going where the light is". The andon is a luminous show activated in its technical application when a problem is found on a workstation to fix it as quickly as possible [45]. It can be caused by an operator or by the equipment where the problem happens automatically. To perform suitable activities, color codes may specify the form or degree of urgency of the anomaly. Initially, it was planned for large production workshops that are very important for visibility. It does, however, refer to other cases, such as call centres, and in its computerized form, in which warning lights can be displayed on the PCs (or mobile devices) of the persons concerned [46].

8. Autonomation or Jidoka:

Jidoka (働化) is an automated shutdown of a machine in the event of detection of a defect. It is a word coined by Sakichi Toyoda in 1896 when he invented the first weaving machine that stops automatically when the yarn breaks; it means "automation with human touch" and has been translated by autonomy (contraction of automation and autonomous) into English; it eliminates the human interference from the machine because if it stops itself, it not required to watch continuously [46].

It has two important concepts in the original TPS:

- One operator can handle many machines at same time, it will improve the human efficiency and save manpower cost.
- To fix them efficiently, the "built-in Quality" identifies quality issues as soon as possible; the full definition also consists of determining the root causes to definitively correct them.

9. Continuous flow:

Unlike batch processing, which consists of producing many products at a time, continuous flow production consists of producing only one product at a time at every stage of the process. It minimizes inventory levels of work in progress and decreases production cycle time, because before going on to the next production stage, each product does not have to wait for others [46].

10. Gemba:

This is undoubtedly one of Lean's most iconic strategies. Gemba, is a Japanese term that means "crime scene" literally. Toyota, which originally used this term, replaced it with the term "Genchi genbutsu" which has a more positive connotation and means "going where the problem is encountered" In fact, the word most widely used today in the industry is the "Gemba walk" usually explained using the Genchi genbutsu translation.

There is a more substantial distinction in theory behind the discrepancies in terms. Whatever the word is, it is a manager's visit to the office. Gemba, however, stresses the inspection and checking of evidence in its original version to make the right decisions. While the "Genchi genbutsu" version, which is like the "management by wandering around" American version, insists more on the casual side and listening to the visited employees [47].

11. Heijunka (Level Scheduling):

Leveling, which means smoothing the preparation or workload in the industrial, is the Heijunka translation. This approach is important to the success of the development of "continuous flow" in practice. It compensates for the fact that orders seldom arrive at a regular pace, in practice [46].

There are two forms of grading:

Volume leveling: the smoothed output produces the average of the orders over a given time, as the orders are of different amounts per day,

leveling by product type: Smoothing is a little more complex, it is a matter of mixing the various items every day according to their processing period to achieve an equal (or nearly identical) average time every day.

The two strategies are merged in practice. The Heijunka box has been developed as a visual medium: it consists of boxes, each representing the type of product (in columns) and the day of the week (in rows), the number of sheets per box being the number of products of the type considered to be manufactured on that day, the sum of the products in the same column being the date of manufacture.

12. Hoshin Kanri:

Hoshin Kanri's literal translation is 'management of the direction'. It means that implementing organizational policy or strategy, or of implementing major improvements, such as restructuring projects, in a wider context. It is the contrary, or rather a supplement to continuing change.

There are three pillars to this method [48]:

A cascaded implementation based on the vision definition: management sets the key directions that are implemented across the organization ("top down" process).

An iterative and participatory process at each hierarchical level: it helps the teams at each level to learn, adjust to reality and appropriate; this process is also called "catchball".

Short and long PDCA cycles: enabling the deployment to be corrected and improved over many time horizons.

13. Plan-Do-Check-Act (PDCA):

The PDCA emerged from a seminar sponsored by the Japanese Union of Scientists and Engineers (JUSE), where W. Edwards Deming updated Shewhart Cycle. It is a method of designing and developing a product according to specifications; it has been introduced by JUSE, and by Kaoru Ishikawa, to be used as a more general method called PDCA. It has become a central component of the Lean theory of quality improvement. It is called the Deming Wheel as well. It is composed of four steps [49]:

Plan: After determining what you want to implement and the targets, plan the actions,

Do: Execute the acts,

Check: Monitor the achievement of acts and goals, understand the outcomes,

Act: Act, apply corrective or enhancement measures [50].

14. Single Minute Exchange of Die (SMED):

It is developed at Toyota by Shigeo Shingo. Its mission is to decrease as much as possible tool changeover times in production [51].

The procedure consists of five stages.

Identify the activities performed: it is important to identify and quantify all activities performed, with waiting times,

Determine inner and outward behaviors:

- Inner activities are relevant to the process of modification that involve the cessation of output.
- Outward activities are performed during the manufacturing or before the manufacturing: component or tool preparation, presetting, etc.

Group external tasks together: Grouping can be eliminating the downtime of output by removing the downtime of processes.

Reduce internal operations time: Detailed analyze can be done for every operation and removed or updated the unwanted things.

Reduce external running time: It has not affect directly but it will increase performance or reduce costs.

15. Standardized Work:

    Operation standardization was invented by Henry Ford, and it is backbone of the TPS. It includes the standardization of systems, tools, operating procedures, and even the extension of parts and components [40, 44].

16. Takt time:

    Takt originates in German and means rhythm.

    It is not a technique strictly talking; it is the basic measurement component of the method of non-stop flow output. This is the manufacturing amount of all item, which in principle essential be equal to the sales price. If all development phases are perfectly balanced at a period equal to Takt time [44] (according to the Heijunka method).

17. Total Productive Maintenance (TPM):

    This technique is based on two main concepts which are included in its name [40, 44].

    - Productive: To perform the maintenance without disturbing the production flow.

    - Total: Contains very variables that influence the correct working of the machines and involves one and all.

    JIPM has established eight TPM pillars [52]:

    i. Independent maintenance: Basic processes carried out by production managers (cleaning, lubrication, inspection, etc.) and the avoidance of breakdowns or the detection of irregularities as early as possible.

    ii. Kobetsu-Kaizen: In the TPS system, it is the equivalent of Kaizen.

    iii. Scheduled maintenance: By preventive work it avoids breakdowns.

    iv. Training and of knowledge management: Trained the technicians and machine operators to improved maintenance.

    v. Maintenance at design stage: In the design of machines or goods, maintenance is considered to promote maintenance processes.

    vi. Quality maintenance: Quality is improved by proper maintenance by removing defects.

    vii. Health, Safety and Environment: This pillar provides workers with a healthy working conditions and support to build a community that attracts equipment consideration.

viii. Office maintenance: Ensuring that the support functions recognize the maintenance issues and, in addition to developing a sense of change of their own processes, can provide support.

18. Value Stream Mapping (VSM):

VSM is the analysis technique that allows all the knowledge flows of a process to be defined and visualized in a synthetic way.

A unreal and visual feature is likely to use of standardized symbols and a definition which, without being exhaustive, must remain at a macroscopic stage [53, 54].

In flow mapping, many pieces of understanding are characteristic:

- The mutual representation of basic and information flows.
- In addition to the other pure development phases, the representation of the journeys and stock phases.
- The identification of key figures for volume for each phase.
- By specifying the processing times and the times between operations, the cumulative time line.
- Identifying the challenges.

19. Waste reduction.

To elimination of waste, which is often more of a Lean concept than a Lean process, is one of Lean core principles. Three forms of waste exist, according to Taichi Ohno [13, 18]:

Muda: Activities with no added value to the finished product; some of its activities, such as quality controls or modifications, are still important [13].

Muri: Tasks that are unnecessary or too difficult [13].

Mura: Variability undergone [13].

1.5 Conclusion

In this study, an overview of the research background has been provided. Eight types of lean waste, nine types of lean manufacturing principles, and nineteen types of lean tools and techniques were identified to eliminate the industrial waste. It concludes that kaizen and 5S are mostly implemented in industries due to no cost or very less cost is required for implementation.

Author details

Shyam Sunder Sharma* and Rahul Khatri
Manipal University Jaipur, Jaipur, Rajasthan, India

*Address all correspondence to: shyamsunder.sharma@jaipur.manipal.edu

References

[1] T. Melton, "The benefits of lean manufacturing: what lean thinking has to offer the process industries," Chemical engineering research and design, vol. 83, no. 6, pp. 662-673, 2005.

[2] A. Brophy, "FT Guide to Lean: How to Streamline Your Organization, Engage Employees and Create a Competitive Edge," Harlow: Pearson Education Limited, 2012.

[3] T. Rotter, C. Plishka, A. Lawal, L. Harrison, N. Sari and D. Goodridge, "What is lean management in health care? development of an operational definition for a cochrane systematic review," Evaluation & the Health Professions, 2018.

[4] J. Womack, D. Jones and D. Ross, "The machine that changed the world," Macmillan publishing company, 1990.

[5] J. Womack and D. Jones, "From Lean Production to the Lean Enterprise," Harvard Business Review, pp. 93-103, 1994.

[6] V. Bernardo, J. A. Garza-Reyes and V. Kumar, "A lean thinking and simulation-based approach for the improvement of routing operations," Industrial Management & Data Systems, 2016.

[7] Ohno, "Toyota Production System: Beyond large-scale production," 1997.

[8] Y. Monden, "Toyota production system: an integrated approach to just-in-time," CRC Press, 2011.

[9] H. Peter and D. Taylor, "Going lean," Cardiff, UK: Lean Enterprise Research Centre Cardiff Business School 1, vol. 528, no. 34, 2000.

[10] A. W. Omran, "Lean production role in improving public service performance in Egypt: challenges and opportunities," Journal of Public Administration and Governance , vol. 4, no. 2, pp. 90-105, 2014.

[11] P. Roberto, "Applying the lessons learned from 27 lean manufacturers.: The relevance of relationships management," International journal of production economics 55.3, vol. 55, no. 3, pp. 223-240, 1998.

[12] Y. Pingyu and Y. yu, "The barriers to SMEs implementation of lean production and its countermeasures– based on SMEs in Wenzhou," International Journal of Innovation, Management and Technology, vol. 1, no. 2, pp. 220-225, 2010.

[13] A. Jamwal, "A study on the barriers to lean manufacturing implementation for small-scale industries in Himachal region (India)," International Journal of Intelligent Enterprise, vol. 6, no. 2-4, pp. 393-407, 2019.

[14] N. Skhmot, "The 8 Wastes of Lean," The Lean way, 2017. [Online].

[15] S. D. Triagus, S. Soeparman and R. Soenoko, "Minimasi waste untuk perbaikan proses produksi kantong kemasan dengan pendekatan Lean Manufacturing," Journal of Engineering and Management in Industrial System, vol. 1, no. 1, 2013.

[16] V. Bernardo, D. Garcia and I. Rosas, "Eliminating transportation waste in food distribution: a case study," Transportation Journal, vol. 48, no. 4, pp. 72-77, 2009.

[17] K. Akhil, "A qualitative study on the barriers of lean manufacturing implementation: An Indian context (Delhi NCR Region)," The International Journal of Engineering & Science, vol. 3, no. 4, pp. 21-28, 2014.

[18] A. K. Tiwari and P. K. Singh, "Study of lean manufacturing, finding its barriers and its relation to quality control: A case study on the manufacturing of crankshaft forging," International Journal For Technological Research In Engineering, vol. 3, no. 8, pp. 1617-1621, 2016.

[19] V. Chahal and M. Narwal, "Impact of lean strategies on different industrial lean wastes," International Journal of Theoretical and Applied Mechanics, vol. 12, no. 2, pp. 275-286, 2017.

[20] L. Teixeira, "Using Lean tools to reduce patient waiting time," Leadership in health services, 2018.

[21] C.-K. Chen, F. Palma and L. Reyes, "Reducing global supply chains' waste of overproduction by using lean principles," International Journal of Quality and Service Sciences, 2019.

[22] P. Arunagiri and A. Gnanavelbabu, "Identification of major lean production waste in automobile industries using weighted average method," Procedia Engineering, vol. 97, pp. 2167-2175, 2014.

[23] A. Dixit, V. Dave and A. P. Singh, "Lean manufacturing: An approach for waste elimination," International Journal of Engineering Research & Technology , vol. 4, no. 4, pp. 532-536, 2017.

[24] S. Vinodh, S. Devarapu and G. Siddhamshetty, "Application of Lean approach for reducing weld defects in a valve component: a case study," International journal of lean six sigma, 2017.

[25] J. Z. Richard and R. D'Angelo, "The Henry Ford Production System: effective reduction of process defects and waste in surgical pathology," American journal of clinical pathology, vol. 128, no. 6, pp. 1015-1022, 2007.

[26] W. O. Aly, "Lean production role in improving public service performance in Egypt: challenges and opportunities," Journal of Public Administration and Governance, vol. 4, no. 2, pp. 90-105, 2014.

[27] T. Jonathan and U. J. M., "The Principles of Lean Manufacturing," Franklin Business & Law Journal , vol. 2016, no. 2 , pp. 57-70, 2016.

[28] A. Leskova, "Principles of lean production to designing manual assembly workststions," Annals of the Faculty of Engineering Hunedoara-International Journal of Engineering, vol. 11, no. 2, pp. 31-36, 2013.

[29] F. M. D. Diego and L. R. Cadavid, "Lean manufacturing measurement: the relationship between lean activities and lean metrics," Estudios gerenciales, vol. 23, no. 105, pp. 69-83, 2007.

[30] Č. Ramunė and M. Vienažindienė, "Lean manufacturing implementation: the main chalenges and barriers," Management theory and studies for rural business and infrastructure development, vol. 35, no. 1, pp. 43-49, 2013.

[31] P. G. D., M. Browaeys and S. Fisser, "Lean and agile: an epistemological reflection," The Learning Organization, 2012.

[32] G. Morteza, "Modeling lean manufacturing success," Journal of Modelling in Management, 2018.

[33] K. Jerry, "Lean principles," Utah Manufacturing Extension Partnership, vol. 68, no. 1, pp. 1-5, 2003.

[34] A. Pereira, "Reconfigurable standardized work in a lean company–a case study," Procedia Cirp, vol. 52, pp. 239-244, 2016.

[35] D. Clark, "Quality improvement in basic histotechnology: the lean approach," Virchows Archiv, vol. 468, no. 1, pp. 5-17, 2016.

[36] M. Lauria and M. Azzalin, "Project and maintainability in the era of Industry 4.0," TECHNE-Journal of Technology for Architecture and Environment, pp. 184-190, 2019.

[37] d.Oliveira and F. Baumont, "Lean Principles in Vertical Farming: A Case Study," Procedia CIRP, vol. 93, pp. 712-717, 2020.

[38] A. P. M., D.-C. José and A. A. Carvalho, "Workplace ergonomics in lean production environments: A literature review," Work, pp. 57-70, 2015.

[39] P. Daryl, E. Alfnes and M. Semini, "The application of lean production control methods within a process-type industry: the case of hydro automotive structures," in IFIP International Conference on Advances in Production Management Systems. Springer, Berlin, Heidelberg, 2009.

[40] S. Sundareshan, D. R. Swamy and T. S. N. Swamy, "A Literature Review on Lean Implementations–A comprehensive summary," International Journal of Engineering Research and Applications, vol. 5, no. 11, pp. 73-81, 2015.

[41] F. Talib, M. Asjad, R. Attri, A. Siddiquee and Z. Khan, "A road map for the implementation of integrated JIT-lean practices in Indian manufacturing industries using the best-worst method approach," Journal of Industrial and Production Engineering, vol. 37, no. 6, pp. 275-291, 2020.

[42] S. Kumar, A. Dhingra and B. Singh, "Lean-Kaizen implementation," Journal of Engineering, Design and Technology , 2018.

[43] N. A. Rahman, A. Sariwati, M. Sharif and M. M. Esa, "Lean manufacturing case study with Kanban system implementation," Procedia Economics and Finance, vol. 7, pp. 174-180, 2013.

[44] K. Salonitis and C. Tsinopoulos, "Drivers and barriers of lean implementation in the Greek manufacturing sector," Procedia Cirp, vol. 57, pp. 189-194, 2016.

[45] J. Hirvonen, "Design and implementation of Andon system for Lean manufacturing," OEV Publication, 2018.

[46] Y. M. Zaki, W. H. W. Mahmood, M. R. Salleh and A. S. M. Yusof, "Review the influence of lean tools and its performance against the index of manufacturing sustainability," International Journal of Agile Systems and Management, vol. 8, no. 2, pp. 116-131, 2015.

[47] A. Cherrafi, S. Elfezazi, B. Hurley, J. A. Garza-Reyes, V. Kumar, A. Anosike and L. Batista, "Green and Lean: a Gemba–Kaizen model for sustainability enhancement," Production Planning & Control, vol. 30, no. 5-6 , pp. 385-399, 2019.

[48] K. R. Kesterson, The Basics of Hoshin Kanri, CRC Press, 2014.

[49] A. Realyvásquez-Vargas, K. C. Arredondo-Soto, T. Carrillo-Gutiérrez and G. Ravelo, "Applying the Plan-Do-Check-Act (PDCA) Cycle to Reduce the Defects in the Manufacturing Industry. A Case Study," Applied Sciences, vol. 8, no. 11 , p. 2181, 2018.

[50] P. M. Patel and V. A. Deshpande, "Application Of Plan-Do-Check-Act Cycle For Quality And Productivity Improvement-A Review," Studies, vol. 2, no. 6, pp. 23-34, 2015.

[51] S. Saad, T. Perera, P. Achanga, E. Shehab, R. Roy and G. Nelder, "Critical success factors for lean implementation within SMEs," Journal of manufacturing technology management , 2006.

[52] M. Dennis and N. Rich, Lean TPM: a blueprint for change, Butterworth-Heinemann, 2015.

[53] I. S. Lasa, C. O. Laburu and R. d. C. Vila, "An evaluation of the value stream mapping tool," Business process management journal, 2008.

[54] N. V. K. Jasti and A. Sharma, "Lean manufacturing implementation using value stream mapping as a tool," International Journal of Lean Six Sigma, 2014.

Chapter 4

Effect of Lean Practices on Organizational Performance

Lokpriya Mohanrao Gaikwad and Vivek K. Sunnapwar

Abstract

The study focuses on the analysis of the direct effect of Lean Manufacturing (LM) practices on operational performance in manufacturing industry. A model for evaluating the effect of LM is developed taking into consideration as a fundamental variable that affects the causal relationship between LM practices and operational performance. A structural equation model was proposed and investigated across the manufacturing industry in India. A structured survey questionnaire was used to collect empirical data from 400 Indian companies. A total of 203 usable responses were obtained giving a response rate of 53%. The data was analyzed using SPSS-AMOS software. The results revealed that LM practices directly and positively affected operational performance. The results indicated that the structural equation model remained invariant across the Industry. The study provides further evidence to managers and practitioner on the effect of LM practices on operational performance in developing countries like India.

Keywords: Lean Manufacturing, Lean practices, organizational performance

1. Introduction

The present powerful market is described by more limited item life cycles and the expanding individualization of items. Along with expanding worldwide rivalry, this puts pressure both on manufacturing organizations' adaptability and on asset effectiveness to satisfy customer need and stay serious [1]. To address these difficulties, manufacturing organizations are compelled to persistently look for new ways to deal with improve their operational performance. Lean manufacturing has over the most recent twenty years seemingly been the most unmistakable approach for improving the operational performance in manufacturing organizations [2, 3]. Based on the straightforward thought of wiping out waste in all forms by focusing in on the exercises that make an incentive for the client [4], it is a low-tech constant improvement approach that centers on representative strengthening and the smoothing out of manufacturing practices. As of late, the innovation situated Industry 4.0 idea is being marked as the following empowering influence of performance improvement.

Manufacturer work in organization to present new plans of action and advances to improve their manageability execution which coordinates the financial, environmental and social responsibilities. Lean manufacturing is a coordinated arrangement of socio-specialized practices planned to consistently dispose of waste to make value and construct a smoothed out, excellent framework [5]. Attributable to the interrelationship among Lean practices, some Lean groups are framed, e.g., just in time (JIT), total quality management (TQM), and human resource management (HRM). They form the basis of Lean creation, every one of which contains a bunch of interrelated and inside steady Lean practices [5, 6]. For instance, JIT incorporates arrangement decrease and little part size. For the most part, manageable execution is worried about a firm's capacity to at the same time consider and balance financial, ecological, and social issues in the conveyance of items or administrations in order to augment esteem [7–9]. It ought to be noticed that practical exhibition in this investigation is characterized as far as its financial and ecological execution measurements. The social performance measurement is excluded. Accordingly, we try to look at if our investigation can discover a route for sustainability minded manufacturer to adjust benefit improvement and natural manageability, which has been at the focal point of consideration among policymakers and the scholarly community [10, 11].

2. Literature review

Lean manufacturing targets reducing waste and non-value added exercises [4]. Inside, underway, this is showed through, in addition to other things, smoothed out, stable, and normalized measures; insignificant inventories; the one-piece stream of items; creation dependent on genuine downstream demand; short setup times; and workers being associated with continuous improvement endeavors [12]. Gaikwad and Sunnapawar [13] opined that if Lean, Green, and Six Sigma strategies help the manufacturing firms to compete in global markets through the impact of sustainable strategy for their business.

Every one of these angles can uphold upgrades in various components of operational performance, for example, item quality and manufacturing cost, lead time, adaptability, and dependability [14]. Since Lean manufacturing was advocated and turned into a standard administration approach, there have been various investigations targeting estimating the real impact of Lean manufacturing on operational performance [15]. Krafcik [16] begat the term Lean and introduced one of the primary examinations to contrast Lean manufactures and common large scale manufacturing firms. Mackelprang and Nair [17] did a meta-examination of 25 articles exploring the connection between Lean practices and execution. While the operationalization of Lean manufacturing rehearses and operational execution will in general shift between examines, the agreement is that the appropriation of Lean manufacturing is emphatically connected with operational execution improvement [17]. Aims of Lean production are to recognize and dispense with the production process wastages for quality improvement, cost decrease, on-time delivery, for example to make effective production processes to confront the most noteworthy rivalry level, so Lean is the most recent device to accomplish it and it getting increasingly remarkable to improve operational and competitive performance [18].

3. Methodology

The empirical data used in this study were collected through a survey distributed to Indian manufacturers that already implemented total quality management practices. The underlying example comprised of all the manufacturing organizations which were on the mailing rundown of an information sharing stage for manufacturing logistics. This underlying example comprised of 400 Indian manufacturing organizations, addressing a wide scope of sectors and company sizes. To the most awesome aspect our insight, the underlying example reflects the Indian business. The link to the survey was disseminated through email, and an aggregate of 212 responses were gathered through an online survey tool. Of these, one of the returned responses needed answers for a few inquiries and was consequently eliminated from the final sample. This examination consequently wound up with a final sample of 203 respondents and a response pace of 53% was noticed.

The study instrument was approved by researching three perspectives: content validity, construct validity, and reliability. To guarantee content validity, a draft survey was pre-tried by two free scholastics with experience in both research project and industry. Also, the survey depended on all around tried and perceived things that have been utilized effectively in different examinations. To evaluate the construct validity, we thought about two viewpoints: convergent validity and discriminant validity [19]. To evaluate convergent validity, we initially examined the unidimensionality of the measures through principal component analysis.

Following the proposals of Carmines and Zeller [20], the things for every one of the constructs were researched independently. For the entirety of the constructs, the Kaiser-Meyer-Olkin measure of sampling adequacy was over the suggested limit of 0.5, and Bartlett's test of sphericity returned p-values beneath 0.001. For all of the autonomous constructs, the items loaded on a single factor, the eigen value surpassed 1.0, the complete difference clarified surpassed half, and all the items' factor loadings were above 0.5, supporting unidimensionality. As added test of convergent validity, the average variance extracted (AVE) and composite reliability (CR) were determined. The recommended thresholds for good convergent validity for these two tests are AVE > 0.5 and CR > 0.7 [21]. For the autonomous factors, the values are over the recommended variables. The dependent variable, operational performance is made out of numerous, unique performance measurements. This implies that the loading factors and thus, AVE and CR will fundamentally be to some degree lower for this construct yet at the same time adequate, as recently proposed by Prajogo and Olhager [22]. To survey discriminant validity, we followed the proposals of Fornell and Larcker [23]. They recommend that to guarantee discriminant validity, the AVE for each construct ought to be more prominent than the square of the construct's bivariate relationships with different constructs. In all cases, this rule was fulfilled. In light of these tests, we expected adequate build legitimacy. To test reliability, the Cronbach's alpha coefficient was determined for every one of the summated scales. All the summated scales have values over the proposed limit of 0.6 Forza [19] and, as needs be, ought to be dependable for additional investigation.

4. Results and discussion

Following **Figure 1** represent the conceptual framework of Lean practices in which Lean practices such as just in time, total productive maintenance, 5S, value stream mapping, single minute exchange of die, etc. plays important role to enhance social, environmental, financial, and operational performance that results overall business excellence in manufacturing industry.

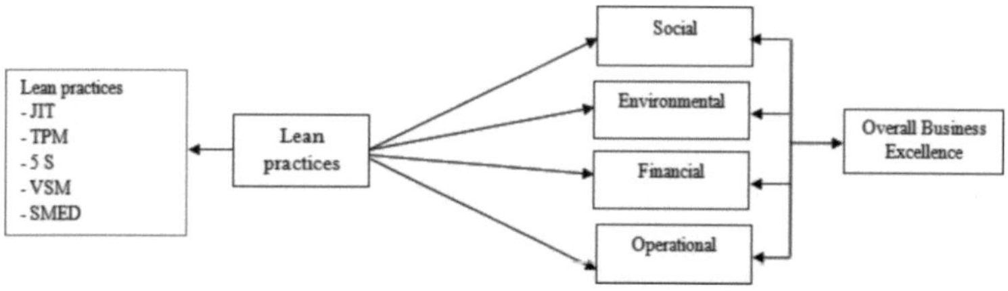

Figure 1.
Conceptual framework of Lean practices.

Structural Equation Model (SEM) for Lean practices and performances:

Figure 2 shows the Structural equation model for Lean practices and its effect on operational, financial, social, and environmental performances.

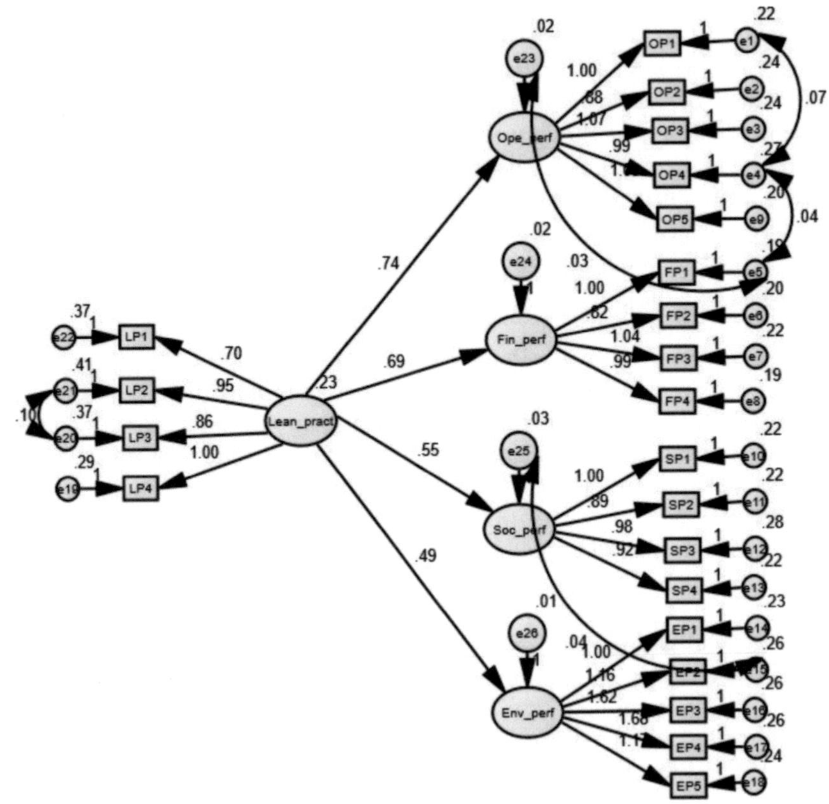

Figure 2.
Structural equation model for Lean practices and performances.

Model Fit Summary
CMIN

Model	NPAR	CMIN	DF	P	CMIN/DF
Default model	53	239.859	200	.028	1.199
Saturated model	253	.000	0		
Independence model	22	1488.573	231	.000	6.444

CMIN/DF = 1.199, in this case less than 3 is good; less than 5 is sometimes permissible [24].
RMR, GFI

Model	RMR	GFI	AGFI	PGFI
Default model	.020	.903	.877	.714
Saturated model	.000	1.000		
Independence model	.116	.333	.269	.304

Goodness of fit indices (GFI) is 0.903, should be higher than 0.9 [24].
Baseline Comparisons

Model	NFI Delta1	RFI rho1	IFI Delta2	TLI rho2	CFI
Default model	.839	.814	.969	.963	.968
Saturated model	1.000		1.000		1.000
Independence model	.000	.000	.000	.000	.000

Comparative fit indices 0.968, (higher than 0.95 great; higher than 0.9 traditional; higher than 0.8 sometimes permissible) [24].
Estimates: Maximum Likelihood Estimates
Regression Weights: (Group number 1 - Default model)

			Estimate	S.E.	C.R.	P	Label
Ope_perf	<---	Lean_pract	.742	.099	7.498	***	
Fin_perf	<---	Lean_pract	.694	.095	7.331	***	
Soc_perf	<---	Lean_pract	.555	.089	6.244	***	
Env_perf	<---	Lean_pract	.493	.086	5.744	***	
OP1	<---	Ope_perf	1.000				
OP2	<---	Ope_perf	.878	.128	6.861	***	
OP3	<---	Ope_perf	1.069	.141	7.599	***	
OP4	<---	Ope_perf	.992	.117	8.460	***	
FP1	<---	Fin_perf	1.000				
FP2	<---	Fin_perf	.821	.126	6.514	***	
FP3	<---	Fin_perf	1.040	.145	7.175	***	
FP4	<---	Fin_perf	.991	.137	7.228	***	
OP5	<---	Ope_perf	1.027	.132	7.769	***	
SP1	<---	Soc_perf	1.000				
SP2	<---	Soc_perf	.893	.162	5.497	***	
SP3	<---	Soc_perf	.976	.179	5.456	***	
SP4	<---	Soc_perf	.919	.164	5.606	***	
EP1	<---	Env_perf	1.000				
EP2	<---	Env_perf	1.160	.225	5.150	***	
EP3	<---	Env_perf	1.619	.279	5.794	***	
EP4	<---	Env_perf	1.676	.286	5.853	***	
EP5	<---	Env_perf	1.175	.224	5.239	***	
LP4	<---	Lean_pract	1.000				
LP3	<---	Lean_pract	.857	.122	7.046	***	
LP2	<---	Lean_pract	.954	.131	7.277	***	
LP1	<---	Lean_pract	.698	.113	6.155	***	

From the above table, it is observed that Lean practices are positively affected on operational, social, environmental, and financial performances ($p \leq 0.05$).

Notes for Model

Computation of degrees of freedom (Default model)

Number of distinct sample moments:	253
Number of distinct parameters to be estimated:	53
Degrees of freedom (253–53):	200

Result

Minimum was achieved

Chi-square = 239.859

Degrees of freedom = 200

Probability level = .028

5. Conclusion

A significant territory to explore is the role Lean manufacturing will play in this new modern period. This examination has reviewed the utilization of various arising advanced innovations just as set up Lean manufacturing practices to explore their relationship with operational performance in manufacturing. It reveals how Lean practices impact sustainable performance. By analyzing data from 203 manufacturing firms, we show that the firm should manage Lean practices in an integrated and coordinated way.

This study adds to explore on manufacturing improvement activities by researching the impact of both Lean manufacturing on operational performance. This examination pointed toward covering the exploration gap with respect to the intelligent impacts of Lean manufacturing on operational execution recently called attention to by Buer, Strandhagen, and Chan [25], just as tending to a portion of the impediments in the prior, comparative investigations. Lean manufacturing has for quite some time been viewed as the 'go-to' answer for improved operational execution and making an improvement culture in the organization. Rinehart, Huxley, and Robertson [26] undoubtedly recommended that Lean manufacturing 'will be the standard production method of the twenty-first century. The operational advantages of utilizing Lean manufacturing have been demonstrated in various past examinations and the aftereffects of the current investigation uphold those discoveries.

The discoveries from the structural equation model confirmed that Lean is as yet an important wellspring of competitive advantage. Albeit large numbers of the thoughts and techniques in Lean manufacturing can be followed far back, the emphasis on making an incentive for the client and decreasing waste are thoughts that will not get old, paying little mind to the mechanical advances that occur.

Acknowledgements

The authors would like to thank the respondents who shared their time and responded to the survey. We would also like to thank the reviewers who helped to improve this chapter.

Disclosure statement

No potential conflict of interest was reported by the author(s).

Author details

Lokpriya Mohanrao Gaikwad[1*] and Vivek K. Sunnapwar[2]

1 Mechanical Engineering Department, SIES Graduate School of Technology, Nerul, Navi Mumbai, India

2 Lokmanya Tilak College of Engineering, Navi Mumbai, India

*Address all correspondence to: lokpriya2007@gmail.com

References

[1] Lasi, H., P. Fettke, H.-G. Kemper, T. Feld, and M. Hoffmann. Industry 4.0. Business & Information Systems Engineering. 2014; 6 (4): 239-242.

[2] Holweg, M. The Genealogy of Lean Production. Journal of Operations Management. *2007;* 25 (2): 420-437.

[3] Found, P., and J. Bicheno. Lean Production. In The Routledge Companion to Lean *Management*, edited by T. H. Netland, and D. J. Powell. 2016; 23-33. New York: Routledge.

[4] Womack, J. P., and D. T. Jones. Lean Thinking: Banish Waste and Create Wealth in Your Corporation. New York: Simon & Schuster. 1996.

[5] Shah, R. and Ward, P.T. Lean manufacturing: Context, practice bundles, and performance. Journal of Operations Management. *2003;* 21(2): 129-149.

[6] Shah, R. and Ward, P.T. Defining and developing measures of Lean production. Journal of Operations Management.2007*;* 25(4): 785-805.

[7] Elkington, J. Cannibals with Forks: The Triple Bottom Line of the 21st Century. New Society Publishers, Stoney Creek. 1998.

[8] Ahi, P., Searcy, C. and Jaber, M.Y. A quantitative approach for assessing sustainability performance of corporations. Ecological Economics.2018*;* 152: 336-346.

[9] Fernando, Y., Jabbour, C.J.C. and Wah, W.X. Pursuing green growth in technology firms through the connections between environmental innovation and sustainable business performance: Does service capability matter?. Resources, Conservation and Recycling. 2019; 141: 8-20.

[10] Akadiri, S.S., Alkawfi, M.M., Uğural, S. and Akadiri, A.C. Towards achieving environmental sustainability target in Italy. The role of energy, real income and globalization. Science of the Total Environment. 2019*;* 671: 1293-1301.

[11] Shahbaz, M., Lahiani, A., Abosedra, S. and Hammoudeh, S. The role of globalization in energy consumption: a quantile cointegrating regression approach. Energy Economics. 2018; 71:161-170.

[12] Chavez, R., W. Yu, M. Jacobs, B. Fynes, F. Wiengarten, and A. Lecuna. Internal Lean Practices and Performance: The Role of Technological Turbulence. International Journal of Production Economics.2015; 160: 157-171.

[13] Lokpriya Gaikwad and Vivek Sunnapwar. An Integrated approach for Lean, Green and Six Sigma strategies: a systematic literature review and future directions for developing a specific framework. TQM Journal. 2020a; 32(2): 201-225. https://doi.org/10.1108/TQM-08-2018-0114, ISSN: 1754-2731.

[14] Marodin, G. A., and T. A. Saurin. Implementing Lean Production Systems: Research Areas and Opportunities for Future Studies. International Journal of Production Research. 2013; 51 (22): 6663-6680.

[15] Ciano, M. P., R. Pozzi, T. Rossi, and F. Strozzi. How IJPR has Addressed 'Lean': A Literature Review Using Bibliometric Tools. International Journal of Production Research.2019; 57 (15-16): 5284-5317.

[16] Krafcik, J. F. Triumph of the Lean Production System. MIT Sloan Management Review. 1988; 30 (1): 41-52.

[17] Mackelprang, A.W., and A. Nair. Relationship Between Just-in-Time Manufacturing Practices and Performance: A Meta-Analytic Investigation. Journal of Operations Management. 2010; 28 (4): 283-302.

[18] Lokpriya Gaikwad, Vivek Sunnapwar. The Role of Lean Manufacturing Practices in Greener Production: A Way to Reach Sustainability. International Journal of Industrial and Manufacturing Systems Engineering. Special Issue: Manufacturing Strategy for Competitiveness. 2020b; 5(1): 1-5. doi: 10.11648/j.ijimse.20200501.11

[19] Forza, C. Survey Research in Operations Management: A Process-Based Perspective. International Journal of Operations & Production Management.2002; 22 (2): 152-194.

[20] Carmines, E. G., and R. A. Zeller. Reliability and Validity Assessment. Thousand Oaks, CA: Sage Publications. 1979.

[21] Hair, J. F., W. C. Black, B. J. Babin, and R. E. Anderson. Multivariate Data Analysis. 7th ed. Upper Saddle River, NJ: Prentice Hall. 2010.

[22] Prajogo, D., and J. Olhager. Supply Chain Integration and Performance: The Effects of Long-Term Relationships. Information Technology and Sharing, and Logistic. 2012.

[23] Fornell, C., and D. F. Larcker. Evaluating Structural Equation Models with Unobservable Variables and Measurement Error. Journal of Marketing Research;1981; 18 (1): 39-50.

[24] Hu, L. T. and Bentler, P.M. Cutoff criteria for fit indexes in covariance structure analysis: conventional criteria versus new alternatives. Structural Equation Modeling: A Multidisciplinary Journal. 1999; 6 (1), pp. 1-55.

[25] Buer, S. V., J. O. Strandhagen, and F. T. S. Chan. The Link Between Industry 4.0 and Lean Manufacturing: Mapping Current Research and Establishing a Research Agenda. International Journal of Production Research. 2018; 56 (8): 2924-2940.

[26] Rinehart, J. W., C. V. Huxley, and D. Robertson. Just Another Car Factory? Lean Production and Its Discontents. Ithaca, NY: Cornell University Press. 1997.

Chapter 5

Enhancement of Textile Supply Chain Performance through Optimal Capacity Planning

Imen Safra and Kaouther Ghachem

Abstract

Manufacturing companies in the textile and apparel field face stiff competition due to the globalization of trade between suppliers, producers and customers. To meet this challenge, they need to be efficient by adopting new lean manufacturing approaches and new analysis and management tools leading to more flexible and agile production and distribution processes. For the textile and apparel industry, where products' life cycle is short due to fashion changes, a new integrated approach of production and distribution planning is needed. Based on linear programming techniques and integrating subcontracting activities, our approach takes into account the characteristics of demand, including its short life cycle, seasonality and fashion effect. For these reasons, a sequential approach is adopted, combining tactical and operational decision levels for production and distribution activities, in order to satisfy customer needs at lower cost by reacting quickly to changes and delivering on time. The deployed approach is structured according to the DMAIC lean tool. Validated on real instances, this approach proves its efficiency by achieving cost reduction when internal production capacity is adequately and efficiently planned.

Keywords: DMAIC lean tool, production-distribution planning, tactical and operational planning, Linear programming, textile and apparel case study

1. Introduction

The success of textile and apparel companies depends largely on supply chain management, which ensures the smooth flow of products to different markets and their availability to customers on time and at the lowest cost. However, this task has become increasingly complex with the expansion of supply chain actors that must be coordinated to ensure a final offer to customers at the desired time and place. There is therefore a need to improve the performance of the supply chain and optimize its management, which requires the simultaneous planning, coordination and management of production and distribution activities to ensure that customer demands are met in a cost effective manner by ensuring the delivery of products on time and at the required location. In this context, lean tools and approaches contribute to the development of the supply chain decision-making process in order to achieve better performance of textile and apparel companies in today's complicated world. That is why, in this chapter, we consider the DMAIC lean approach and we focus on the integration of production and distribution operations managed by a textile and apparel supply chain manufacturer.

Our choice of the textile and apparel sector corresponds well to the problematic we are studying of a production chain with multiple actors geographically spread all over the world. The activities of these actors must be optimized in order to determine the adequate offer of each unit of the production chain. Moreover, the nature of the textile product, which is not a homogeneous good but highly diversified, short-lived and subject to the effects of fashion as described in the bibliographical references [1–7]. For these products, we distinguish two types of orders: (1) preseason orders (PO) that include products for the next season with a medium delivery time and (2) replenishment orders (RO) that include products for the current demand season with a short delivery time. In addition, the textile and apparel industry is highly competitive worldwide and is rapidly changing due to the complexity of demand, which is subject to the effects of fashion and marketing. This results in changes on the supply side, with some businesses disappearing and others expanding depending on the degree of rapid reaction to demand and customer tastes, as an inadequate response to demand can result in unsold inventory and lost sales opportunities.

We considered in this work a planning approach integrating tactical and operational decision levels and taking into account textile and apparel industry specificities. Using a rolling horizon, the proposed approach identifies the quantities to be produced, stored and delivered while minimizing the total cost of production and distribution. Production flexibility is ensured by the consideration of low-cost overseas subcontractors to whom standard products with predictable demand can be assigned. Local subcontracting and overtime are short-term solutions to deal with the unpredictability of demand related to ROs at the operational level. This work is structured according to the DMAIC approach and will be detailed accordingly while defining the specifics of each phase.

2. Phase "define"

As detailed by [8], the 'define' phase of the DMAIC methodology presents a definition of the problem and what the customer requires. Hence, it is the backbone of a successful project. The define phase starts with clarifying the problem statement and analyzing the customer requirements and ensures that the project goals are aligned to these requirements.

2.1 The problem statement

Facing a worrying decline in market share for textile-apparel manufacturers in the context of the competitive battle, these manufacturers must act by creating new offers combining low prices, reduced lead times and improved services. This can be ensured by carrying out adequate resource planning at different levels of decision making and coordinating activities associated with the various stakeholders in the chain. Moreover, in regard to more selective consumer behavior, the emergence of customized products with short life cycles and taking into account the different types of orders, manufacturers must satisfy customers by being reactive, fast and more and more flexible while offering a better quality, price and lead time performance. In this context, a coordinated control of flows between suppliers, producers and customers can only lead to a fast, personalized and optimal response, in accordance with the expectations of end consumers. Traditionally, the various supply chain actors manage their resources independently, and the planning and management of production and distribution resources is done with little or no coordination. This decentralized management can lead to additional costs due to the placement of unexpected and urgent orders at subcontractors' units or by scheduling costly overtime. On the other hand, additional delays may be caused by re-planned resources and the delayed arrival of a few productions due to the arrival of these urgent orders to be placed as soon as possible. Similarly, large inventories and long product cycles may occur as soon as the producer opts for large production quantities to anticipate demand for the entire season, not to mention the risk of increasing unsold stock.

Our work focuses on a global approach that integrates production and distribution activities. The related literature review is presented by [9–13]. In this work, we are interested in addressing the problem of production-distribution coordination applied to the textile and apparel field. A presentation of the different types of coordination at the supply chain level and a review of the literature dealing with this aspect are detailed by [14, 15].

Indeed, most studies on integrated production and distribution have focused on products for which demand is stable because they have a long-life cycle [16, 17]. But this is not the case for apparel products that have a short life cycle and whose demand can only be accurately estimated once the product is on the shelf once the season has started. Similarly, few production planning models have taken into account the flexibility of production capacities. However, our models provide this flexibility through outsourcing and overtime [18, 19]. Therefore, it is necessary to adapt production and distribution planning models to the reality of textile and apparel supply chains in order to optimize them, taking into account the unpredictable and unforeseen aspect of demand while aiming to reduce production and distribution lead times to better match production to demand. In this way, production can be flexible and can be adapted to market needs. Thus, it is necessary to define production and distribution planning models that take into account the specificity of the apparel supply chain. The objective of this study is to start filling this gap.

It applies to the case of a large Tunisian textile company (see **Figure 1**) that owns several units of apparel production and two warehouses located in Tunisia. It may also use outsourcing with local or overseas subcontractors in China to meet part of its demand. The company adopts a business model of delivery commitment. It commits to a delivery date for any order received and is responsible for shipping costs. Finished products are stored in warehouses until they are delivered on time to customers. The transportation modes used are trucks, ships and airplanes. The transportation cost includes fixed and variable fees. Each product has a production set-up cost and a variable cost.

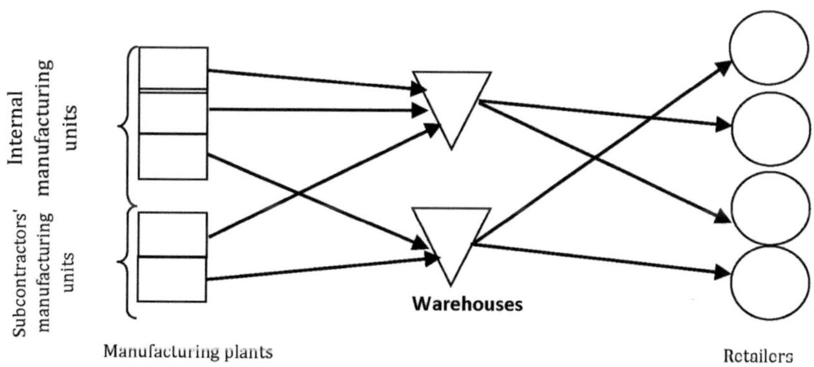

Figure 1.
The textile and apparel supply chain.

Received orders, over a season, cover a large number of product references. The number of product groups exceeds 100. The company receives two types of orders from local and overseas retailers: POs and ROs. POs, which have a lead time of several months, are planned and scheduled to satisfy the following season's collections. However, urgent ROs, which have shorter delivery times, must be produced to fill retailer shortages or to replace unsold inventory. Due to changing fashion trends and the short life cycles of textile and apparel products, historical data alone cannot accurately predict next season sales [20]. Moreover, it is very difficult to forecast specific customer needs for apparel products, leading retailers to use in-season replenishment after revising their forecasts based on demand observed in the first few weeks of the current season. Therefore, it is a periodic process of adjusting retailers' sales forecasts for different products taking into account new information from recent sales.

2.2 Challenges of the proposed planning approach

As detailed in **Figure 2**, the proposed approach is based on an integrated production-distribution planning at two decision levels while considering the specificities of the apparel supply chain. Thus, the approach involves decisions at the tactical and operational level and takes into consideration both POs and ROs. Also, the approach considers flexibility of production capacity to ensure a better match between supply and demand. We consider at the level of operational planning overtime and subcontracting activities to accommodate the internal capacity shortage caused by fluctuations of demand and short lead time of customer orders. The main goal of the current study is to reduce overall supply chain costs by approximately 10%.

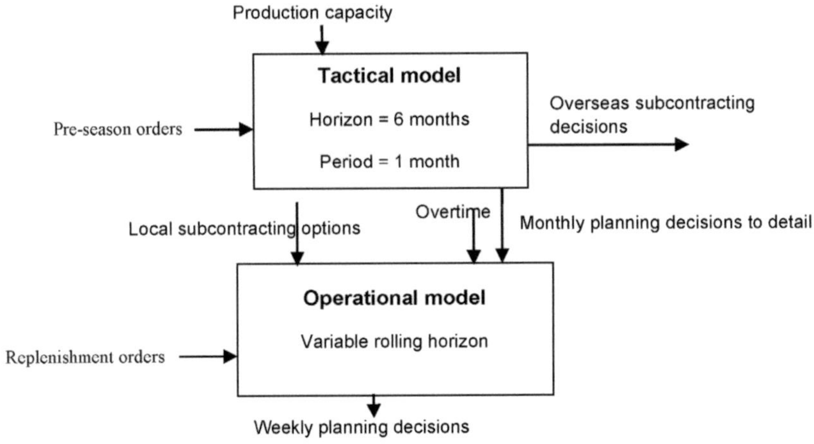

Figure 2.
Hierarchical planning approach.

2.3 The process definition

The definition of the high-level process map gives the team an eye of bird's view about the project. One of the most used high-level map is the SIPOC which details the Supplier, Input, Process, Output and Customers. By completing the SIPOC, the detailed map (Swimlane Map) can be developed after a series of Gemba walk and several discussions with the teamwork.

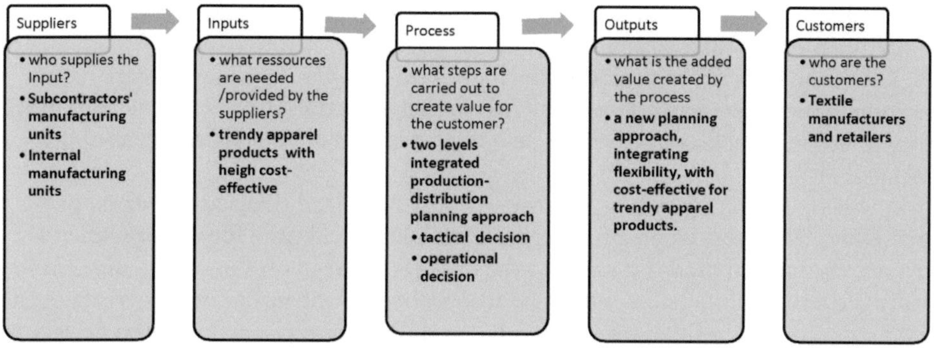

Figure 3.
SIPOC diagram.

The process definition of the SIPOC diagram, as detailed in **Figure 3**, ties the different steps of the new proposal solution to create the added value. In the current case, a new approach based on two level integrated production-distribution plan and composed of two level of strategies (tactic and operational) is developed. Knowing that textile and apparel manufacturers are currently dealing with unpredictable and short-term ROs when production for the next season is already ongoing. The company uses two types of sub-contractors: - overseas specialized subcontractors who offer products at very low prices but with long delivery times, and - local capacity subcontractors that the company uses in case of production saturation but who offer prices 20% higher than the internal production costs. In addition to this flexibility provided by subcontracting, the company can resort to overtime with higher costs than production in regular hours.

3. Phase "measure"

In this section, we will define the current state in order to analyze it and to identify the gap between the actual and the desired situations. To do this, we will structure this part in three phases. First of all, we will detail the developed measurement system. Then, we will detail our data collection plan and our experimental data. Finally, we will identify our desired situation and the gap with the current one.

3.1 The measurement system analysis

3.1.1 Approach description

Our measurement system aims to define our sequential production and distribution planning approach while evaluating the current situation of the company. The objective is to satisfy POs and ROs within the required deadlines, while minimizing the production, subcontracting, capacity under-utilization, storage and distribution costs.

Our approach is based on two mathematical models that are developed at two different decision levels [21, 22]. The first model focuses on a tactical level of a 6-month horizon with a monthly periodicity and decides on pre-season quantities to be placed internally and with overseas subcontractors with long lead times. Each time a new order arrives, it is inserted with a rolling horizon in order to be placed optimally on the available resources.

The second operational model considers a monthly horizon with a variable periodicity between 8 and 11 weeks. This model is used to detail tactical confirmed quantities over weeks and to insert new urgent orders arriving through a rolling horizon.

At the operational level, urgent orders with short lead times are inserted progressively. However, when a new order with a long lead time for the pre-season arrives, the tactical model is run and the order is inserted on the rolling horizon to study the possibility of subcontracting it to overseas subcontractors. Thereafter, the production will be refined over weeks using the operational model, if the decision taken at the tactical level affects production internally. This operation is repeated until all orders for the season have been placed.

Our approach results in a production, storage and distribution plan that takes into account the assignments of overseas subcontractors and the assignments of new orders that arrive at the operational level. The latter are detailed by week during the first 2 months. Given the principle of the rolling horizon, decisions taken during the first week are fixed and the related costs are recorded. However, decisions taken in following periods will be revised once the model has been run in the second week.

3.1.2 Mathematical formulation

3.1.2.1 Tactical planning model

As detailed in appendix A, at the tactical level, the model decides on: - monthly quantities to be manufactured internally and at subcontractors, – monthly quantities to be stored, – monthly quantities to be distributed per period, taking into account the different modes of transport. The objective is to minimize the total cost of production, storage and distribution. It should be noted that for this current situation, the parameter α_{kt} considered in Eq. (3) and Eq. (6) is equal to 100%. Means that we are considering all the available production capacity at the tactical level.

3.1.2.2 Operational planning model

As shown in **Figure 4**, the operational horizon ranges from 8 to 11 weeks. The length of this horizon is defined according to the position of the first week in the month.

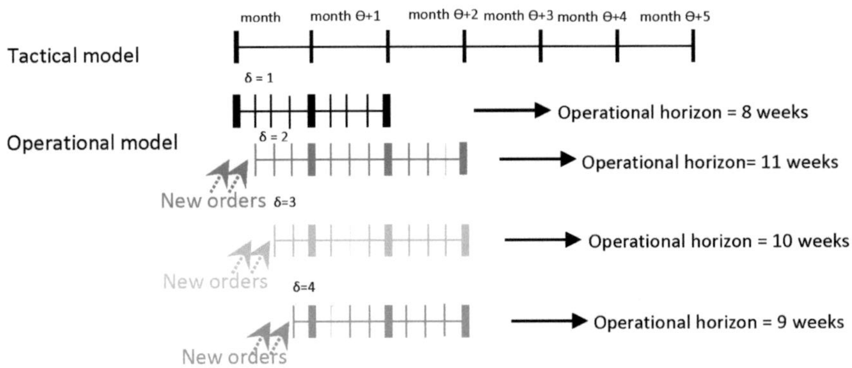

Figure 4.
Variable operational planning horizon.

For example, if the first week of the planning horizon corresponds to the second week of the month Θ, then the length of the planning horizon is set to 11 weeks because tactical decisions related to the month Θ + 2 must also be taken into account.

We denote by a couple (t, s) the weeks in the operational planning model, where s is the position of the week in month t. Operational planning includes a set of periods $TS_{\Theta\delta}$ and takes place at the beginning of week δ of month Θ (as detailed in appendix B).

For the operational planning model, detailed in appendix C, the length of the planning horizon is justified by the fact that POs resulting from tactical planning should be reliably detailed at the operational level. Hence, to properly place decisions made at the tactical level, the operational planning horizon must reach the end of the month.

3.2 Data collection plan

Based on company reports translating historical data, Gemba walks and after meetings and team benchmarking, we were able to collect the necessary information to set the required parameters for the proposed models.

Before detailing experimental data, it is important to identify the established planning assumptions:

- Demand is dynamic and deterministic,
- Storage cost is defined according to the average level of storage between the beginning and the end of the period,
- Under-utilization capacity cost is estimated according to the average hourly cost of labor/machine.
- In the operational model, we consider only local subcontractors.
- Overtime is considered to allow greater flexibility when managing the unpredictability and short lead time of ROs.

3.2.1 Experimental data

The relevant company delivers about 200 references of products to 30 retailers per year through 3 knitting manufacturing plants located in Tunisia. Products are transferred to customers through two local warehouses storing finished products ordered by local and overseas retailers. These warehouses are characterized by their limited storage capacity and a storage cost of approximately 5% of the unit production cost per unit.

The shipments can be carried out by trucks, for local deliveries, or by aircraft and by ships, for overseas connections. Our mathematical models decide on the mode of transport to be adopted according to the delivery times involved. Indeed, a delay of at least 5 weeks is necessary to deliver the products by ship. However, aircraft shipments are made within the same week. The considered transportation costs are composed of fixed costs, depending on the number of shipments made, and variable costs depending on the shipped quantities.

Considering the 200 variety of manufactured products, the internal production costs vary from 3 to 35 euros. In order to accommodate the limited internal capacity, flexibility is ensured by scheduling overtime. However, overtime activity is limited to 25% of production capacity after regular working hours and costs 40% more. The internal flexibility is reinforced by a subcontracting activity with 10 local subcontractors and one overseas one located in China. The local subcontractors have enough capacity to meet the ordered quantities and fill the limited capacity of the internal production sites. The latter offer products at prices 20% higher than the cost of internal production. As for the Chinese subcontractor, it can manufacture large volumes of products but with delivery times exceeding 2 months. The latter offers basic products at costs that are about half the internal production cost. Our planning models decide on production allocations based on available capacities and assigned lead times. The overall focus is to meet customer orders on time and at lower cost.

Our proposed approach is run over 6 months, generating a weekly production schedule identifying the quantities to be manufactured, stored and distributed. The proposed models are solved using the package ILOG OPL Studio V6.3/ Cplex 11 and are run on an intel Core i5 PC with a 2.3 Ghz processor and 512 MB of memory. The planning model at the tactical level takes into account approximately 122,000 constraints and 66,000 variables, of which more than 5,000 are binary. However, the operational model contains 55,000 constraints and 3,000 binary variables among the 25,000 considered variables in the model. An almost optimal solution, with a deviation of 10^{-4} from optimality, is obtained for all the executed models.

3.3 Current situation and the gap with the desired one

Our approach evaluates the current situation of the apparel manufacturer who incurs a logistic cost equal to 2864 k€ obtained for the 6 months.

In order to improve the situation, we aim at considering additional flexibility at the tactical planning level in order to better accommodate the unpredictable orders that will be placed at the operational level. A decrease of the overall logistic cost is expected.

4. Phase "analyze"

At the end of the six-month simulations, we obtain a weekly production, storage and distribution schedule for the various products, taking into account the tactical model's assignments and the unforeseen and urgent demands that arrive at the operational level. The cumulative costs obtained for the first few weeks of the operational model applied on a rolling horizon, added to the tactical cost of production at overseas' subcontractor, obtained by the tactical model, represent the total cost of production, storage and distribution for the six months. This cost, as reported in **Table 1**, is evaluated to 2 864 K€.

Period	Cost (K€)	Period	Cost (K€)	Period	Cost (K€)
Overseas sub Mars	320	May S 1	92	July S 1	86
March S 1	79	May S 2	69	July S 2	90
March S 2	99	May S 3	93	July S 3	105
March S 3	84	May S 4	92	July S 4	91
March S 4	98	Overseas sub June	223	August S 1	113
April S 1	83	June S 1	83	August S 2	106
April S 2	83	June S 2	90	August S 3	93
April S 3	87	June S 3	89	August S 4	108
April S 4	76	June S 4	218	Total cost	**2864**

Table 1.
Weekly logistics costs.

	Quantities	%
Internal manufacturer's production	259359	90,5
Internal overtime production	1833	0,6
Subcontractor's production	25507	8,9
Total produced quantities	286699	100,0

Table 2.
Production assignment

Weekly costs represent the sum of production costs in regular hours, overtime and at local subcontractors added to the costs of storage, underutilization of internal production capacity and product deliveries.

Obtained results for the current situation show that production is mainly affected to Internal manufacturing units 90.5%. However, 8.9% are affected to overseas subcontractors as shown in **Table 2**.

Based on these obtained results, we notice, on one hand, a production allocation that leads to an overload of internal production capacity and costly overtime. On the other hand, some productions get started in overtime, especially when the internal production capacity over regular hours is partially used. This is mainly due to the due dates position of the POs through the month. Indeed, since the productions planned in the internal units over a month are detailed at the operational level by week, it seems mandatory in some cases to massively produce during the first weeks of the month to meet the predefined delivery dates. Consequently, it is necessary to produce in overtime some products that the production capacity during regular hours cannot meet. Meanwhile, for the remaining weeks of the month, the capacity of the internal production is under-utilized. In this case, the ROs, which arrive at an operational level, are assigned to the subcontractors since the internal capacity of production is fully used by the production of the pre-season items decided at the tactical level.

This obtained cost seems to be too high because decisions at the tactical level are made without taking into account what may arrive at the operational level as urgent and unforeseen ROs. This cost can be improved to be more competitive in the market through greater flexibility at a tactical level. This flexibility could positively affect the allocation of orders that arrive at an operational level.

To analyze the current situation of the textile and apparel supply chain and try to identify the root causes of the performance decrease in this field, we establish the following Ishikawa diagram shown in **Figure 5**.

Based on this analysis, we confirm that it is necessary to reduce lead times through better resource management and better planning that will reduce the operational workload on operators. Taking into account the specificities of this sector and the requirements imposed by markets and customers, it is essential to adjust to needs, as soon as they are presented, through a better flexibility of resources at a tactical level.

The 5 P tool (**Figure 6**) is also used to identify the root cause of the problem so that the required actions can be taken to improve performance.

It is quite clear now that the solution is to provide some flexibility at the tactical planning and not to allocate rigidly anticipated productions without allowing sufficient flexibility to place the orders that arrive at the operational level.

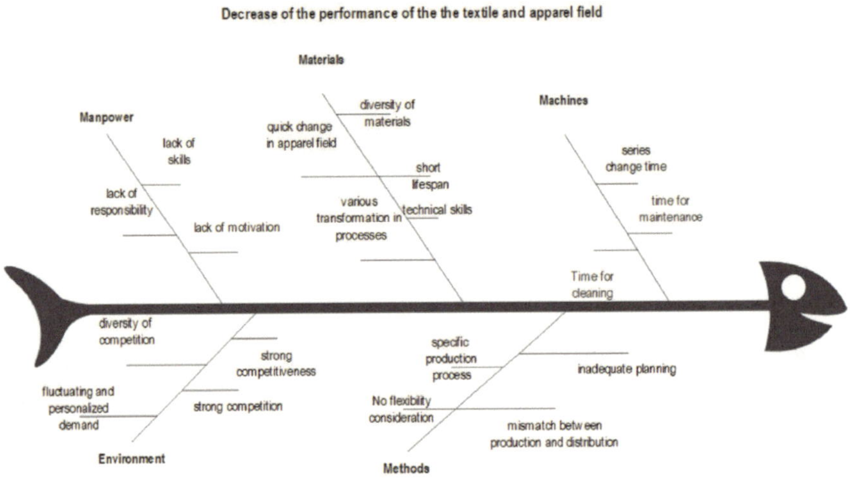

Figure 5.
Ishikawa diagram.

Figure 6.
5P tool.

5. Phase "improve"

To solve the root cause, at this phase, we have to introduce more flexibility at the tactical level by considering only a percentage of the production capacity. The other part of the production capacity is considered as reserve capacity. Thus, it can be used only at the operational level to efficiently meet RO with short due dates without disrupting the ongoing production. During this study, we need to achieve one main objective: how to satisfy the retailers' pre-season in addition to the ROs that must be in time. This objective will be reached by minimizing the internal capacity underutilization, storage, distribution operation and also the overall supply chain cost incurred by internal production, subcontracting.

The availability of products based on ROs during the season is risky for the retailer since it largely depends on the flexibility, responsiveness and efficiency of the suppliers involved. Therefore, to meet retailer orders and ensure deliveries on-time, production flexibility becomes crucial and a key competitive issue for any textile or apparel manufacturing company.

At the tactical model a reserve production capacity (RPC) is considered. We denote the percentage of internal production capacity that can be used to fulfill POs by $\boldsymbol{\alpha}$. α_{kt} is the reserve related to an internal site k over a period t. As it can be noted, $(100 - \alpha_{kt})$ represents the percentage of internal capacity reserved to fulfill in-season ROs.

Meaning that the parameter α_{kt} considered in Eqs. (3) and (6) is less than 100%.

At the operational level, the RPC considered at the tactical level is released and the entire internal capacity can be used in addition to overtime. This will provide more flexibility to accommodate unforeseen and urgent ROs.

Let us now consider the operational level, the RPC considered at the tactical level is released. In addition, all internal capacity can be used to overtime. This will result in greater flexibility to respond to unexpected and urgent ROs.

5.1 The reserve production capacity estimation

The impact of considering the RPC at the tactical planning level on supply chain costs is investigated. During this experimentation, the same value for this RPC for all internal manufacturing units is used. Firstly, for each month of the six-month tactical planning horizon, a fixed RPC is considered. The percentage of the available internal production capacity for PO planning is therefore a fixed value (α). Secondly, a RPC with monthly variation is considered. The percentage of internal generation capacity at the tactical planning model level is therefore a value that varies monthly and is noted (αt), with t indexed to the month.

The RPC needs to be estimated. Afterwards, the available two-years historical demand data is used to estimate the RPC (1-α or 1-αt). Thus, it is obtained by calculating the ratio: reserve production/total internal production during regular hours. The resulting internal production capacity rates are shown in **Table 3**.

Month →	M1	M2	M3	M4	M5	M6	Average
Rate year N-2 (%)	62	90	89	71	61	76	75
Rate year N-1 (%)	75	91	72	91	87	95	85
Average rate	69	91	81	81	74	85	80

Table 3.
Observed internal production capacity rates used based on 2-year historical demand data

Hereinafter, different values of the RPC are tested. The objective is twofold. The first one is to underline the importance of integrating RPC into tactical planning to improve flexibility. The second one is to emphasize the need to develop adequate methods based on historical demand data and can provide an efficient estimation of the RPC.

5.2 Production and distribution planning using a fixed reserve production capacity

Different values of α are tested. These values vary from 70% to 100% with a difference of 5% between two consecutive values. The curve depicting the variation in supply chain cost as a function of α is plotted in **Figure 7**. The curve is characterized by an almost convex shape. In addition, for α values equal to 70%, 75% and 100% higher costs are observed. Indeed, the reserve of 30 to 25% of production capacity for ROs leads to the allocation of many orders to subcontractors at the tactical level. Consequently, a significant underutilization of capacity is observed at the operational planning level. If no reserve capacity is being considered at the tactical planning level (which is the current practice in the company), We note that, at the operational planning level, many ROs are assigned to subcontractors or produced during overtime as internal production capacity is used during regular working hours to accommodate POs.

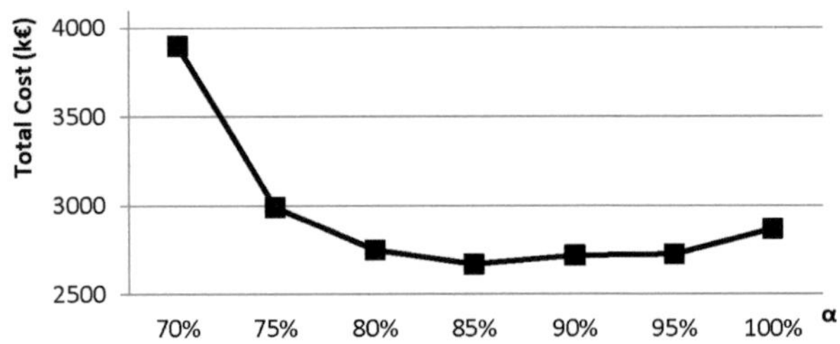

Figure 7.
Supply chain cost variation with α.

Note that in the considered real case study, the optimal supply chain cost is reached at a value α around 85%. Hence, a RPC of about 15% ensures a production flexibility that minimizes the cost of the supply chain. The average value obtained from the historical database (presented in **Table 3**) is equal to α = 80%. The cost of the corresponding supply chain is equal to 2,746 k€. This translates into a saving of 4% compared to current practice (α = 100%). When the proposed planning approach is used with α equal to 85%, the cost saving over current practice is equal to 7%.

5.3 Production and distribution planning with a variable monthly reserve production capacity

In this section, it is proposed to evaluate the monthly variable RPC. For each month t of year N, we take for each year N - 1 and N - 2 the average of the percentage of internal production capacity used as the value of αt (represented in **Table 3**). A supply chain cost equal to 2575 k€ is obtained by introducing the values of αt into the tactical planning model and sequentially applying the tactical and operational models. The cost of the supply chain is, as observed, less than that obtained when considering a fixed RPC equal to 20%. This method used to estimate the RPC leads to a 6% cost reduction compared to the previously used method. Furthermore, there is a saving of 10% compared to current practice (**Figure 8**).

This cost saving resulted from allocating six months of production to internal manufacturing units and subcontractors, as illustrated in **Figure 9**.

Firstly, when considering a monthly variable RPC at the tactical level, there is a better use of internal production capacity. Second, we find that some production is performed during overtime when the internal production capacity is not fully used during regular hours.

The reason for this can be explained by the position of PO due dates within the month. Since production in the internal manufacturing units over 1 month from tactical planning is detailed per week at the operational level, massive production in the first weeks of the month is sometimes necessary to meet the delivery due dates.

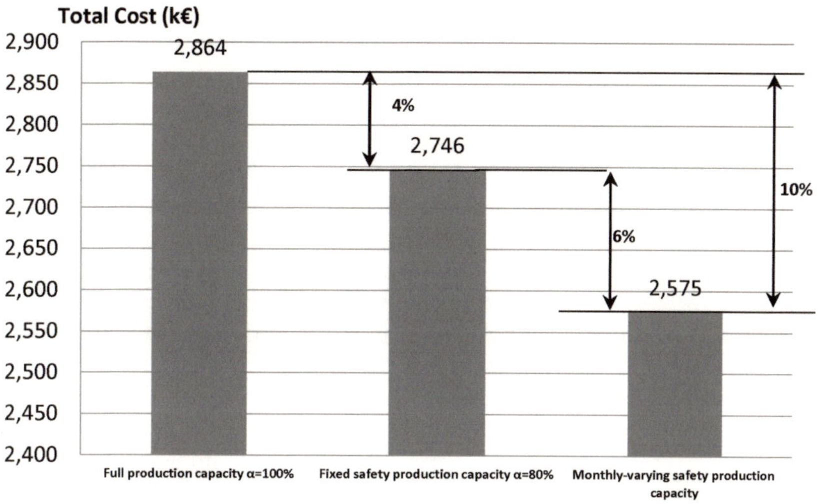

Figure 8.
Supply chain cost comparison.

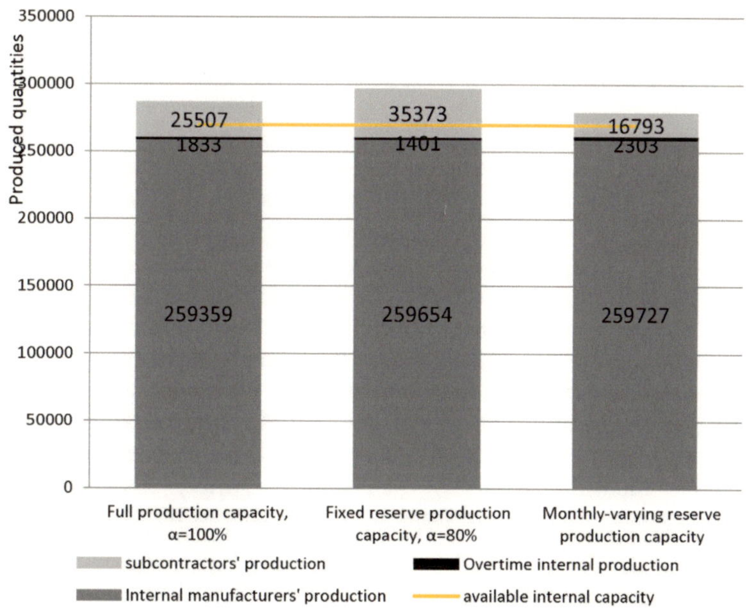

Figure 9.
Production assignment.

As a result, overtime is needed as production during regular hours cannot reach the requested quantities. Internal production capacity for the remaining weeks of the month may be underused.

When we consider a fixed RPC (for α = 80% and α = 100%), the quantities produced at subcontractors' manufacturing units are bigger than those performed when a variable monthly RPC is considered. Consequently, the allocation of production to subcontractors is better optimized for a monthly variable RPC.

Considering this result, we emphasize the importance of the monthly variable RPC. This reserve is adjusted to ROs by assigning, at the tactical level, some productions to subcontractors while maintaining sufficient and accurate internal production capacity at the operational level to appropriately handle ROs.

However, the quantities produced in the subcontractors' manufacturing units are particularly high when we consider, at the tactical level, a fixed α, equal to 80%.

Meanwhile, the total produced quantities over the 6 months are higher than those produced when considering α equal to 100%, or a monthly variable RPC. This is due to the demand monthly variation. Actually, when we consider a fixed α equal to 80%, two situations can arise. Firstly, the ROs to be satisfied during the month require more than the available capacity and consequently more than the RPC considered. In such a case, the subcontracting activity is the main solution. Second, ROs to be placed during the month require less than the available capacity and so less RPC. Therefore, ROs to be filled for next few weeks are processed in advance to minimize capacity under-utilization. When α is equal to 100%, ROs are assigned to subcontractors since internal production capacity is overloaded by POs.

As conclusion, the use of a variable RPC at the tactical level, allows efficient use of internal production capacity and optimizes the allocation of production to subcontractors. Nevertheless, the performance of capacity planning can be improved if more accurate and reliable historical demand data is used and if forecasting methods for predicting the monthly variable RPC are carried out.

By studying the three cases mentioned above, we underline the important effect of taking into account a suitably defined RPC on the supply chain cost.

6. Phase "control"

Due to intense competition, variable economic and environmental conditions, changing wage rates and fluctuating oil prices, the control phase of the DMAIC methodology will be focused on establishing the changes and standardizing the results given in the previous phases. Consequently, sensitivity analysis is chosen to assess the effect of changes in demand and variations in subcontracting and transportation costs on the supply chain cost. Three parameters are examined in this sensitivity analysis: demand, transportation cost and subcontracting cost to assess their impact on planning decisions and supply chain performance.

During our experimentation, fifteen scenarios are considered by varying (1) the demand, (2) the cost of transport and (3) the cost of subcontracting between −50% to +50% of their current values. By considering the three scenarios (α = 100%, monthly fixed α, monthly varying α), the cost of the supply chain is calculated for each case.

6.1 Sensitivity analysis of demand

A 50% increase in demand leads to an increase in supply chain costs, as explained in **Table 4**. Nevertheless, if we consider a RPC, a decrease in this total cost is recorded. For all considered scenarios, the best cost is obtained when a monthly variable RPC is considered.

	D−50%	D−20%	D	D+20%	D+50%
α = 100% (k€)	1658	2152	2864	3285	3794
α = 80% (k€)	1632	2020	2669	3067	3581
α variable (k€)	1596	1981	2575	2972	3389

Table 4.
Cost variation according to demand

These results confirm the importance of our approach and encourage the idea of using RPC to reduce supply chain costs. We note a saving of 4% compared to current practice when demand is reduced by half. This proves that the use of a monthly variable RPC yields better results. If demand is increased by half, the gain is 11%. The proposed approach becomes essential when demand is relatively high. If demand is low, the internal capacity will accommodate the demand without any additional costs. As a result, the RPC becomes less important and will avoid situations of under capacity due to urgent orders arriving at the operational level.

6.2 Sensitivity analysis of transportation and subcontracting costs

Identical trends are observed in transport and subcontracting costs savings are also obtained when we consider a RPC at the tactical planning level (**Figure 10**). It is worth noting that the greatest savings are achieved when considering a monthly variable RPC. Moreover, savings become more important with increases in these two costs. Lower transportation costs lead to the outsourcing of some internal production to overseas subcontractor's manufacturing unit, as the latter offers very competitive prices, especially for most basic products. Subsequently, at the tactical planning level, some internal production capacity is unused; therefore, enforcing a RPC is meaningless. For this reason, the lowest savings are observed when transport costs are halved. Nevertheless, outsourcing abroad is no longer the preferred option when transport costs increase. This promotes the use of a RPC to avoid the use of full production capacity at the tactical level. Internal production (regular and overtime) and locally subcontracting are the adequate options to cover capacity requirements.

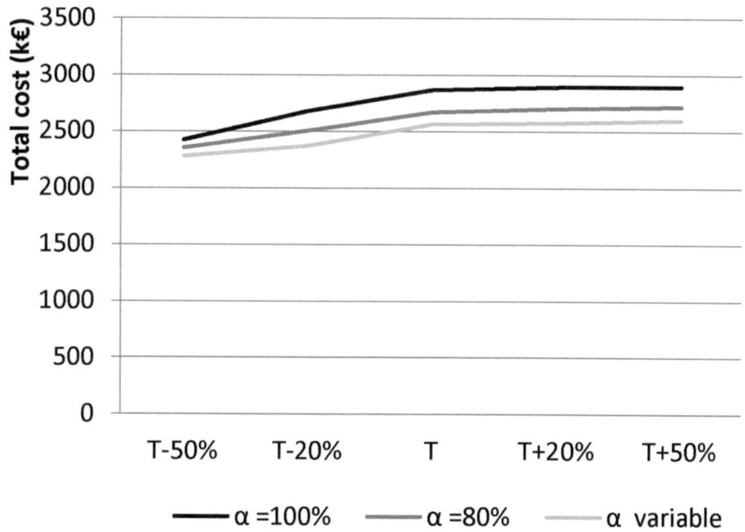

Figure 10.
Transportation cost variation.

When the cost of subcontracting is reduced by half, this activity is more profitable than the internal production. In this case, part of the internal production is manufactured at local subcontractors. In addition, the under-utilization cost prevents the full transfer of quantities to local subcontractors' plants. Internal production capacity is currently under-utilized; however, this situation results in lower supply chain costs due to lower production costs. In this case, the consideration of a RPC is no longer significant.

Nevertheless, increased supply chain costs are noted when the cost of subcontracting increases, especially for α = 100% (current situation). In practice, subcontracting is prevented until internal production capacity is completely used; thus leading to local subcontracting at higher costs (**Figure 11**). In these situations, reserve capacity appears to be a substantial consideration in ensuring the use of foreign subcontracting (which is cheaper than local subcontracting) at the tactical level. If the cost of subcontracting is increased by half, taking into account a RPC that varies on a monthly basis makes it possible to achieve a cost savings of 11% regards current practice.

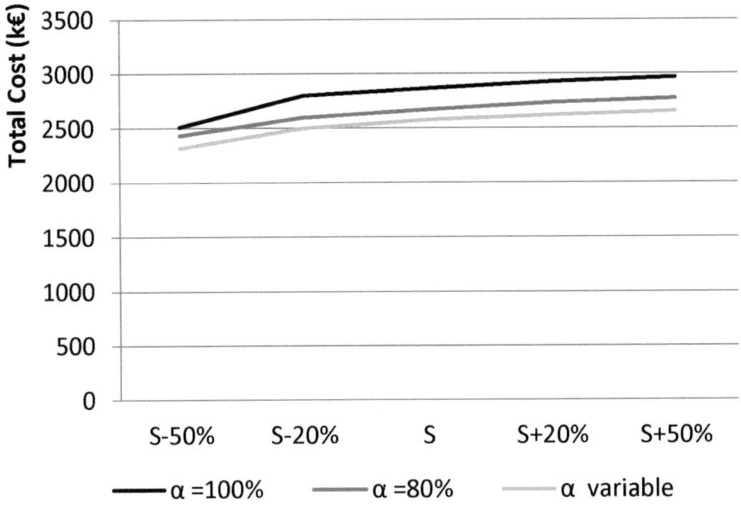

Figure 11.
Subcontracting cost variation.

The sensitivity analysis confirms the interest of our approach taking into account a RPC. Indeed, when demand or transportation or subcontracting costs increase, our approach allows us to adequately place urgent and unpredictable orders that arrive at the operational level at the lowest cost.

This approach provides a decision tool for textile and apparel manufacturers who are constantly faced with two types of orders: long lead time orders dedicated for the next season and urgent and unpredictable orders that are related to the current season. Moreover, this approach is applied to any type of industry where there are two types of customers: - premium customers with short lead time orders, – classic customers with long lead time orders. Indeed, through flexibility and responsiveness to needs, our approach will be able to place orders in the right location and at the lowest cost, taking into consideration a RPC and a rolling horizon. This will guarantee customer satisfaction that will gain in competitiveness, on both cost and lead time aspects, in today's highly competitive environment.

Furthermore, taking into account the type of information introduced, the performance of the supply chain can be improved. The more reliable this information is, the better the performance. Indeed, it is important for the customer to share his sales information with the producer so that the latter can prepare in advance, using adequate forecasting methods, to best accommodate these orders. In this case, the estimate of the RPC can be adjusted for more reliability and flexibility in order to forecast future orders that may come in. This is one of the perspectives and ideas to be explored for this work.

7. Conclusions

In this chapter, the DMAIC methodology was chosen and applied to perform a complicated problem of a textile company. Our aim is to satisfy customer needs at lower cost while ensuring prompt and punctual deliveries. To achieve this, a sequential approach integrating tactical and operational decisions for textile and apparel supply chain planning has been implemented with an emphasis on the flexibility provided by the consideration of RPC at the tactical level. As a result, newly arrived urgent orders, with short lead times, can be placed optimally at production sites, via the rolling horizon.

During the definition phase of the DMAIC methodology, we have defined the problem statement and presented the proposed planning approach. Then, we established the SIPOC diagram in order to identify the different steps of our approach which ensures flexibility of production and distribution activities' planning considering textile and apparel sector specificities: fashion effect, demand fluctuations.

In the first step, we have detailed our measurement system analysis by introducing the two mathematical models used to evaluate the performance of the current situation of the apparel company, taking into account the full available production capacity. Next, we presented our data collection plan by describing the experimental data that were collected. Finally, we outlined the desired situation taking into account additional flexibility at the tactical level.

In the "analysis" phase, we presented the obtained results when assessing the current situation by detailing production assignments over different locations. We also performed an extended analysis using the Ishikawa diagram and the 5P tool in order to underline the interest of our approach.

During the "improve" phase, we outlined the Improvements achieved in the current situation. To do so, we started by testing different RPC values in order to identify the optimal one to be taken into account at the tactical level. Then, we evaluated the performance of our approach by considering a fixed RPC then a monthly variable one. Finally, we evaluated the efficiency of our approach to optimally respond to urgent orders arriving at the operational level. Our approach is evaluated over a six-month planning horizon, but it remains applicable over longer planning horizons.

"Control" phase is devoted to sensitivity analysis while studying the effect of some parameters' variation on the cost of the supply chain. The three considered parameters are: demand, transport cost and subcontracting cost. The main focus of this section is to prove the interest of our approach to place, at the lowest cost, urgent orders that arrive at the operational level, even when demand and cost increase. For a better performance of the considered supply chain, the importance of cooperation between the manufacturer and the retailers, based on information sharing, was also emphasized.

Appendix A. The tactical planning model

In model formulation, we consider the following sets and indices, parameters, and decision variables.

Sets and indices:

K: set of manufacturing units $k \in K$; $K = U \cup V$.

U: set of internal manufacturing units, $k \in U$.

V: set of subcontractors' manufacturing units, $k \in V$.

I: set of retailers, $i \in I$.

J: set of warehouses, $j \in J$.

P: set of products, $p \in P$.

L: set of transportation modes, L = {trucks, ships, aircraft}, $l \in L$.

T: set of periods included in the planning horizon, $t \in [1 .. |T|]$.

Parameters:

In the tactical model, the parameter (D_{pit}) expresses the need of retailer i, in product p, to serve during period t. During period t, the quantities to be delivered are manufactured in production sites k characterized by a limited capacity (U_{kt}). Production incurs variable and fixed production costs per product per period (C_{pkt}, S_{pkt}) or subcontracting costs (G_{pkt}). A monthly cost of under-utilization of internal production capacity (CSU_{kt}) is also considered to penalize the non-utilization of available internal resources. Each product is characterized by a production lead time (Tp_p) and a product unit volume (V_p). The quantities manufactured are then transported to warehouses and incur inventory holding costs (KP_{pjt}). The warehouses are characterized by a limited storage capacity (W_j). The delivery lead time is noted by (e_l). Each means of transport has limited capacity (Cap_l). Variable and fixed distribution costs from sites to warehouses (CT_{kjplt}, CF_{kjplt}) and from warehouses to retailers (CS_{jiplt}, CFS_{jiplt}) are also addressed.

Decision variables:

$Z1_{kjplt}$: quantity of product p to deliver, via transportation mode l, from manufacturing unit k to warehouse j over period t,

$Z2_{jiplt}$: quantity of product p to deliver, via transportation mode l, from warehouse j to retailer i over period t,

X_{pkt}: produced quantity, of product p, in manufacturing unit k over period t.

SU_{kt}: unused production capacity at internal manufacturing unit k over period t.

J_{pjt}: inventory level of product p in warehouse j at the end of period t.

Y_{pkt} =1 if product p is produced in manufacturing unit k over period t; 0 otherwise.

$N1_{kjlt}$: transported quantity from manufacturing unit k to warehouse j over period t by use of transportation mode l.

$N2_{jilt}$: transported quantity from warehouse j to retailer i over period t by use of transportation mode l.

Model formulation (M1)

The tactical production–distribution planning model is formulated as an ILP that aims at minimizing the overall cost in the considered supply chain network.

$$\begin{aligned}
\text{Min}\Bigg(& \textstyle\sum_{t\in T}\sum_{p\in P}\sum_{k\in U} C_{pkt}X_{pkt} + \sum_{t\in T}\sum_{p\in P}\sum_{k\in U} S_{pkt}Y_{pkt} + \displaystyle\sum_{t\in T}\sum_{p\in P}\sum_{k\in U} G_{pkt}X_{pkt} \\
& + \sum_{k\in U}\sum_{t\in T} CSU_{kt}SU_{kt} \sum_{j\in J}\sum_{t\in T}\sum_{p\in P} KP_{pjt}\left(J_{pjt-1} + J_{pjt}\right)/2 \\
& + \sum_{t\in T}\sum_{p\in P}\sum_{k\in K}\sum_{l\in L}\sum_{j\in J} CT_{kjplt} * V_p * Z1_{kjplt} + \sum_{t\in T}\sum_{p\in P}\sum_{k\in K}\sum_{l\in L}\sum_{j\in J} CS_{jiplt} * V_p * Z2_{jiplt} \\
& + \sum_{t\in T}\sum_{p\in P}\sum_{k\in K}\sum_{l\in L}\sum_{j\in J} CF_{kjplt} * V_p * Z2_{jiplt} + \sum_{t\in T}\sum_{p\in P}\sum_{k\in K}\sum_{l\in L}\sum_{j\in J} CF_{kjplt} * N1_{kjlt} \\
& + \sum_{t\in T}\sum_{p\in P}\sum_{k\in K}\sum_{l\in L}\sum_{j\in J} CFS_{jilt} * N2_{jilt} \Bigg)
\end{aligned}$$

Subject to

$$J_{pjt} = J_{pjt-1} + \sum_{l\in L}\sum_{k\in K} Z1_{kjplt-e_l} - \sum_{l\in L}\sum_{k\in K} Z2_{kjplt},\ j\in J; p\in P; t\in T \text{ and } t \geq e_l \tag{1}$$

$$\sum_{p\in P} J_{pjt} \leq W_j,\ j\in J \text{ and } t\in T \tag{2}$$

$$\sum_{p\in P} Tp_p * X_{PKT} \leq \alpha_{kt} * V_{kt}, k\in K \text{ and } t\in T \tag{3}$$

$$X_{pkt} \leq M * Y_{pkt}, k \in K, p \in P and t \in T \tag{4}$$

$$Y_{pkt} \leq X_{pkt}, k \in V, p \in P and t \in T \tag{5}$$

$$SV_{kt} \geq \alpha_{kt} * V_{kt} - \sum_{p \in P} Tp_p * X_{pkt}, k \in K and t \in T \tag{6}$$

$$X_{pkt} = \sum_{j \in J} \sum_{l \in L} Z1_{kjplt}, k \in K and t \in T \tag{7}$$

$$D_{pit} = \sum_{j \in J} \sum_{l \in L} Z2_{jiplt-e_l}, i \in I, p \in P, t \in T, t \geq e_l \tag{8}$$

$$\sum_p V_p * Z2_{jiplt} \leq N2_{jilt} * Cap_l, j \in J, i \in I, l \in L and t \in T \tag{9}$$

$$\sum_p V_p * Z1_{kjplt} \leq N1_{kjlt} * Cap_l, j \in J, k \in K, l \in L and t \in T \tag{10}$$

$$Y_{pkt} \in \{0, 1\} k \in K, p \in P and t \in T \tag{11}$$

$$\begin{aligned} &Z1_{kjplt} \in \mathbb{N}, Z2_{jiplt} \in \mathbb{N}, X_{pkt} \in \mathbb{N}, J_{pjt} \in \mathbb{N}, N1_{kjplt} \in \mathbb{N}, \\ &N2_{jiplt} \in \mathbb{N}, SV_{kt} \in \mathbb{N} \;\; k \in K, j \in J, p \in P, t \in T, l \in L and i \in I \end{aligned} \tag{12}$$

The objective function aims at minimizing the total cost composed of set-up cost, variable production cost, subcontracting cost, internal capacity underutilization cost, inventory holding cost, transportation costs from manufacturing units to warehouses and transportation cost from warehouses to retailers. Transportation costs are composed of variable and fixed costs. The first, depends on quantity to deliver using transportation mode. While the second is proportional to the number of shipments.

The constraints (1) determine the stock level of product p in warehouse j at the end of period t. Constraints (2) guarantee that over each period, the total stored quantity is limited by the storage capacity. Constraints (3) ensure that the produced quantities do not exceed production capacity. Constraints (4) and (5) establish the relationship between binary and integer variables. Constraints (6) with the objective function identify the underutilized internal production capacity. Constraints (7) guarantee the delivery of all produced quantities to warehouses. Constraints (8) ensure that delivered products from warehouses to retailers meet on time demand. Constraints (9) and (10) guarantee the respect of transportation capacity. Constraints (11) and (12) are integrity constraints.

B. The set of periods in the operational planning model

The set of periods in the operational planning model used at week δ of month Θ is presented at the table below. For example, to construct an operational planning at the beginning of the second week (δ = 2) of month Θ, the periods involved are (Θ,2), (Θ,3), (Θ,4), (Θ + 1,1), (Θ + 1,2), (Θ + 1,3), (Θ + 1,4), (Θ + 2,1), (Θ + 2,2), (Θ + 2,3), and (Θ + 2,4) and they are listed in the third column of table below ($TS_{\Theta 2}$).

Set of periods in the operational planning model used at week δ of month Θ.

	δ = 1	**δ = 2**	**δ = 3**	**δ = 4**
(t, s) in	(Θ,1)	(Θ,2)	(Θ,3)	(Θ,4)
	(Θ,2)	(Θ,3)	(Θ,4)	(O + 1,1)
	(Θ,3)	(Θ,4)	(Θ + 1,1)	(Θ + 1,2)
$TS_{\Theta\delta}$	(Θ,4)	(Θ + 1,1)	(Θ + 1,2)	(Θ + 1,3)

δ = 1	δ = 2	δ = 3	δ = 4
(Θ + 1,1)	(Θ + 1,2)	(Θ + 1,3)	(Θ + 1,4)
(Θ + 1,2)	(Θ + 1,3)	(Θ + 1,4)	(Θ + 2,1)
(Θ + 1,3)	(Θ + 1,4)	(Θ + 2,1)	(Θ + 2,2)
(Θ + 1,4)	(Θ + 2,1)	(Θ + 2,2)	(Θ + 2,3)
	(Θ + 2,2)	(Θ + 2,3)	(Θ + 2,4)
	(Θ + 2,3)	(Θ + 2,4)	
	(Θ + 2,4)		

C. The operational planning model

The same predetermined parameters of the tactical model are maintained for the operational planning model except few adjustments. Since the tactical and operational models consider different periods, w has been added here to the parameters and decision variables to indicate that they are related to a one-week period. The operational planning model determines the weekly quantities to be produced, stored and delivered $(t,s) \in TS_{\Theta\delta}$. It is worth knowing that the production plans obtained from the tactical model, for month t such as $(t,s) \in TS_{\Theta\delta}$, represent inputs to be considered at the operational level and must be weekly detailed.

In addition to the notation introduced in the tactical planning model, we consider the following two parameters and two decision variables related to overtime:

Parameters:

UHw: overtime production capacity in internal manufacturing unit $k \in U$ at week s of month t with (t,s) in $TS_{\Theta\delta}$.

CHw_{pkts}: overtime production cost in internal manufacturing unit $k \in U$ at week s of month t with (t,s) in $TS_{\Theta\delta}$.

Decision variables

XHw_{pkts}: quantity of product p produced during overtime in internal manufacturing unit $k \in U$ at week s of month t with (t,s) in $TS_{\Theta\delta}$.

$YHw_{pkts} = 1$ if there is production of p during overtime in internal manufacturing unit k at week s of month t; 0 otherwise with (t,s) in $TS_{\Theta\delta}$.

Model formulation (M2)

The main objective is to minimize the overall cost composed of: -weekly production cost, – weekly set-up cost during regular working hours and overtime, – weekly subcontracting cost, – weekly internal production capacity underutilization cost, – weekly holding inventory cost, – weekly variable and fixed transportation costs from manufacturing units to warehouses and from warehouses to retailers.

$$
\begin{aligned}
\text{Min}\Bigg(& \sum_{(t,s)\in TS_{\theta\delta}} \sum_{p\in P} \sum_{k\in V} Cw_{pkts} Xw_{pkts} + \sum_{(t,s)\in TS_{\theta\delta}} \sum_{p\in P} \sum_{k\in V} Sw_{pkts}\left(Yw_{pkts} + YHw_{pkts}\right) \\
& + \sum_{(t,s)\in TS_{\theta\delta}} \sum_{p\in P} \sum_{k\in V} Gw_{pkts} Xw_{pkts} + \sum_{(t,s)\in TS_{\theta\delta}} \sum_{p\in P} \sum_{k\in V} CHw_{pkts} XH_{pkts} \\
& + \sum_{(t,s)\in TS_{\theta\delta}} \sum_{k\epsilon V} CSVw_{kts} SVw_{kts} + \sum_{j\in J} \sum_{(t,s)\in TS_{\theta\delta}} \sum_{p\in P} KPw_{pjts}\left(Jw_{pjts-1} + Jw_{pjts}\right)/2 \\
& + \sum_{(t,s)\in TS_{\theta\delta}} \sum_{p\in P} \sum_{k\in K} \sum_{l\in l} \sum_{j\in J} CTw_{kjplts} V_p Z1w_{kjplts} \\
& + \sum_{(t,s)\in TS_{\theta\delta}} \sum_{p\in P} \sum_{i\in I} \sum_{l\in l} \sum_{j\in J} CSw_{jiplts} V_p Z2w_{jiplts} \\
& + \sum_{(t,s)\in TS_{\theta\delta}} \sum_{p\in P} \sum_{l\in L} \sum_{k\in K} \sum_{j\in J} CFw_{kjplts} N1w_{kjplts} \\
& + \sum_{(t,s)\in TS_{\theta\delta}} \sum_{l\in L} \sum_{p\in P} \sum_{i\in I} \sum_{j\in J} CFSw_{kjplts} N2w_{kjplts} \Bigg)
\end{aligned}
$$

Constraints (1), (2), and (8)–(12) of the tactical model are also included at the operational level while introducing weekly parameters and decision variables. They ensure the balance of production flows, the respect of storage capacity, the satisfaction of retailer demand, the respect of transportation capacity [(9) and (10)], and guarantee the integrity of the decision variables [(11) and (12)]. Constraints (3)–(7) are changed to incorporate full production capacity and overtime as follows:

$$\sum_{p\in P} Tp_p * XHw_{pkts} \leq VHw_{kts}, k \in \mathcal{U}; (t,s) \in TS_{\theta\delta} \quad (13)$$

$$\sum_{p\in P} Tp_p * Xw_{pkts} \leq Vw_{kts}, k \in \mathcal{K}; (t,s) \in TS_{\theta\delta} \quad (14)$$

$$XHw_{pkts} \leq M * \left(YHw_{pkts} + Yw_{pkts}\right), k \in \mathcal{U}; p \in \mathcal{P}; (t,s) \in TS_{\theta\delta} \quad (15)$$

$$Xw_{pkts} \leq M * \left(YHw_{pkts} + Yw_{pkts}\right), k \in \mathcal{K}; p \in \mathcal{P}; (t,s) \in TS_{\theta\delta} \quad (16)$$

$$YHw_{pkts} + Yw_{pkts} \leq 1, k \in \mathcal{U}; p \in \mathcal{P}; (t,s) \in TS_{\theta\delta} \quad (17)$$

$$YHw_{pkts} \leq XHw_{pkts}, k \in \mathcal{U}; p \in \mathcal{P}; (t,s) \in TS_{\theta\delta} \quad (18)$$

$$Yw_{pkts} \leq Xw_{pkts}, k \in \mathcal{K}; p \in \mathcal{P}; (t,s) \in TS_{\theta\delta} \quad (19)$$

$$SVw_{kts} \geq Vw_{kts} - \sum_{p\in P} Tp_p * Xw_{pkts}, k \in \mathcal{U};\ (t,s) \in TS_{\theta\delta} \quad (20)$$

$$XHw_{pkts} + Xw_{pkts} = \sum_{l\in L}\sum_{j\in J} Z1w_{kjp\ lts}, k \in \mathcal{K};\ p \in \mathcal{P};\ (t,s) \in TS_{\theta\delta} \quad (21)$$

Constraints (13) and (14) guarantee the respect of the production capacity in regular working hours and on overtime. Constraints (15), (17) and (18) ensure that the cost of overtime is only taken into account if the same products are not previously produced. Constraints (16) and (19) establish the relationship between binary and integer variables. Constraints (20) with the objective function set the underutilized internal production capacity. Constraints (21) ensure that all production quantities are delivered to the warehouses.

Constraints (22)–(26) are also considered at the operational model:

$$\sum\nolimits_{(t,s)\in TS_{\theta\delta}/\delta\geq 1} Xw_{pkts} = X_{pk\theta} - \sum\nolimits_{s=1}^{\delta-1} Xw_{pk\theta s}, k \in \mathcal{K}; p \in \mathcal{P}; t = \Theta \quad (22)$$

$$\sum\nolimits_{(t,s)\in TS_{\theta\delta}/\delta\geq 1} Xw_{pkts} = X_{pk\theta+1}, k \in \mathcal{K}; p \in \mathcal{P}; t = \Theta + 1 \quad (23)$$

$$\sum\nolimits_{(t,s)\in TS_{\theta\delta}/\delta\geq 1} Xw_{pkts} = X_{pk\theta+2}, k \in \mathcal{K}; p \in \mathcal{P}; t = \Theta + 2 \quad (24)$$

$$YHw_{pkts} \in \{0,1\}, k \in \mathcal{K};\ p \in \mathcal{P};\ (t,s) \in TS_{\theta\delta} \quad (25)$$

$$YHw_{pkts} \in \mathbb{N}, k \in \mathcal{K};\ p \in \mathcal{P};\ (t,s) \in TS_{\theta\delta} \quad (26)$$

Constraints (22), (23) and (24) guarantee coherence with the tactical decisions made. Finally, constraints (25) and (26) ensure the integrity of the new decision variables.

Author details

Imen Safra[1,2*] and Kaouther Ghachem[3]

1 LACS, National Engineering School of Tunis, University of Tunis El Manar, Tunis, Tunisia

2 National Engineering School of Bizerte, University of Carthage, Bizerte, Tunisia

3 College of Engineering, Industrial and Systems Department, Princess Nourah Bint Abdulrahman University, Riyadh, Saudi Arabia

*Address all correspondence to: imen.safra@centraliens.net

References

[1] Choi TM, Yeung WK, Cheng TCE, Yue X. Optimal Scheduling, Coordination and the Value of RFID Technology in Garment Manufacturing Supply Chains. IEEE Transactions on Engineering Management. 2018; 65(1): 72–84. DOI: 10.1109/TEM. 2017.2739799.

[2] De Toni A, Meneghetti A. The production planning process for a network of firms in the textile-apparel industry. International Journal of Production Economics. 2000; 65(1): 17–32. DOI: 10.1016/S0925-5273(99) 00087-0.

[3] Karacapilidis NI, Pappis CP. Production planning and control in textile industry: A case study. Computers in Industry. 1996; 30 (2): 127–144. DOI: 10.1016/0166-3615(96) 00038-3.

[4] Margaret B, Lucy D. Forecasting Adding value: challenges for UK apparel supply chain management - a review. Production Planning & Control: The Management of Operation. 2011; 22 (3): 210–220. DOI: 10.1080/ 09537287.2010.498574.

[5] Pan A, Leung SYS, Moon KL, Yeung KW. Optimal reorder decision-making in the agent-based apparel supply chain. Expert Systems with Applications. 2009; 36: 8571–8581. DOI: 10.1016/j.eswa.2008.10.081.

[6] Shen B, Chan HL, Chow PS, Thoney-Barletta KA. Inventory management research for the fashion industry. Int. J. Inventory Research. 2016; 3(4): 297–317. DOI: 10.1504/IJIR.2016.082325.

[7] Silva C, Magalhaes JM. Heuristic lot size scheduling on unrelated parallel machines with applications in the textile industry. Computers & Industrial Engineering. 2006; 50 (1–2): 76–89. DOI: 10.1016/j.cie.2006.01.001.

[8] Henderson KM, Evans JR. Successful Implementation of Six Sigma: Benchmarking General Electric Company. Benchmarking: An International Journal. 2000; 7: 260–282. DOI: 10.1108/14635770010378909.

[9] Chen ZL. Integrated Production and Distribution Operations: Taxonomy, Models, and Review. Chapter 17 of the book "Handbook of Quantitative Supply Chain Analysis: Modeling in the E-Business Era", edited by Simchi-Levi D, Wu SD, Shen ZJ, Kluwer Academic Publishers. 2004.

[10] Chen ZL. Integrated Production and Outbound Distribution Scheduling: Review and Extensions. Operations Research. 2010; 58 (1): 130–148. DOI: 10.1287/opre.1080.0688.

[11] Erengüç SS, Simpson NC, Vakharia AJ. Integrated production/ distribution planning in supply chains: An invited review. European Journal of Operational Research. 1999; 115 (2): 219–236. DOI: 10.1016/S0377-2217(98) 90299-5.

[12] Mula J, Peidro D, Díaz-Madroñero M, Vicens E. Mathematical programming models for supply chain production and transport planning. European Journal of Operational Research. 2010; 204: 377–390. DOI: 10.1016/j.ejor.2009.09.008.

[13] Sarmiento AM, Nagi R. A review of integrated analysis of production-distribution systems. IIE Transactions. 1999; 31: 1061–1074. DOI: 10.1023/A: 1007623508610.

[14] Arshinder Kanda A, Deshmukh SG. Supply chain coordination: Perspectives, empirical studies and research directions. International Journal of Production Economics. 2008; 115(2): 316–335. DOI: 10.1016/j.ijpe.2008. 05.011.

[15] Thomas DJ, Griffin PM. Coordinated Supply Chain Management. European Journal of Operation Research. 1996; 94(1): 1–15. DOI: 10.1016/0377-2217(96)00098-7.

[16] Sousa R, Shah N, Papageorgiou LG. Supply chain design and multilevel planning-An industrial case. Computers and Chemical Engineering. 2008; 32 (11): 2643–2663. DOI: 10.1016/j.compchemeng.2007.09.005.

[17] Tsiakis P, Papageorgiou LG. Optimal production allocation and distribution supply chain networks. International Journal of Production Economies. 2007; 111(2): 468–483. DOI : 10.1016/j.ijpe.2007.02.035.

[18] Jolayemi JK, Olorunniwo FO. A deterministic model for planning production quantities in a multi-plant, multi-warehouse environment with extensible capacities. International Journal of Production Economics. 2004; 87(2): 99–113. DOI: 10.1016/S0925-5273(03)00095-1.

[19] Selim H, Araz C, Ozkarahan I. Collaborative production-distribution planning in supply chain: fuzzy goal programming approach. Transportation Research Part E: Logistics and Transportation Review. 2008; 44(3): 396–419. DOI: 10.1016/j.tre.2006.11.001.

[20] Mostard J, Teunter R, Koster MBM. Forecasting demand for single-period products: A case study in the apparel industry. European Journal of Operational Research. 2011; 211: 139–147.

[21] Safra I, Jebali A, Jemai Z, Bouchriha H, Ghaffari A. Capacity planning in textile and apparel supply chains. IMA Journal of Management Mathematics. 2019 ; 30(2): 209–233. DOI: 10.1093/imaman/dpy006.

[22] Safra I. Vers une approche intégrée de planification de la production et de la distribution : cas de l'industrie textile [thesis]. Ecole centrale de Paris; 2013.

Chapter 6

From Lean Manufacturing to Lean Construction: How Principles, Tools, and Techniques Evolved

Sevilay Demirkesen

Abstract

Lean manufacturing first emerged in the automotive industry. However, low productivity and low efficiency in production are major problems for the majority of industries relying on a heavy workforce. Being one of these, the construction industry suffers from low productivity rates along with inefficient work practices. To prevent those, the industry has shifted its focus from the traditional approach to a more innovative one, which is called Lean construction. Lean construction aims to maximize value while minimizing waste. Therefore, it intends to create safer, smoother, and more efficient processes to eliminate waste. This chapter focuses on Lean construction and highlights the generic Lean tools and techniques practiced in the construction industry indicating its historical journey from Lean manufacturing. The chapter aims to raise awareness towards the efficiency of Lean methods in the construction industry with respect to practices observed in manufacturing.

Keywords: lean manufacturing, productivity, efficiency, lean construction, lean methods

1. Introduction

The foundation of Lean thinking dates back to the 1900s, when Henry Ford, founder of Ford Motor Company, came up with an entire production process relying on interchangeable parts with standard work and moving conveyance for creating a flow production [1]. Melton [2] defines Lean as a revolution indicating that Lean is not just utilizing tools and techniques or making a few changes in processes, rather he defines Lean as a complete change in businesses to observe supply chain operations, managerial decisions, and daily work of employees in an organization. The authors of the book named "The Machine that Changed the World", which is one of the most influential books implied that the Lean way results in better products at a lower cost as well as encouraging employees to overcome challenges in production processes [3]. Even though Lean manufacturing has first found its roots at Ford, it was later investigated by Toyota Motor Company. The Japanese engineer Taiichi Ohno, who had several visits to Ford factories to observe production processes. However, Taiichi Ohno found some methods implemented at Ford as needing improvements. Therefore, Sakichi Toyoda, his son, Kiichiro Toyoda, and Taiichi Ohno came up with the concept of Lean Manufacturing, which was first called just-in-time (JIT) production [4]. Taiichi Ohno was responsible for implementing the new ideas that evolved into the Toyota Production System (TPS). Then, Taiichi Ohno hired Shigeo Shingo to work on the setup reduction problem at Toyota. Shingo later named this successful process the famous Single Minute Exchange of Dies (SMED) system. This is how production ideas evolved at Toyota leading to technical innovations.

The lean manufacturing concept was first articulated as a shop floor practice to reach higher efficiency in processes being implemented with JIT and Toyota Production System (TPS) [5, 6]. It was also mentioned that Lean manufacturing in the 1980s rather focused on shop floor techniques and inventory reduction as well as value-added processes in the supply chain [7, 8]. Lean manufacturing is now implemented as a popular manufacturing practice in various countries and industries [9]. The ultimate goal intended by Lean organizations is to have a high-quality organization responsive to customer demands with no waste. On the other hand, most manufacturing organizations fail to realize the transformation for Lean due to a range of challenges faced [6]. The majority of the previous studies implied that even though most Lean organizations aim to implement Lean in the best way, they fail at some point as a matter of fact [10, 11]. However, the organizations are still seeking ways to improve their Lean approach and effectively practice Lean methods.

The success of Lean thinking in the manufacturing industry positively affected the construction industry. However, the construction industry is a conservative and fragmented industry, which makes innovations less welcomed by industry practitioners [12]. On the other hand, low productivity rates and intentions to improve workforce efficiency led the construction industry to implement innovative technologies.

The term 'Lean Construction' was first articulated by the International Group for Lean Construction (IGLC) in 1993. Glenn Ballard and Greg Howell, the two construction practitioners who first considered Lean in construction projects, started the Lean Construction Institute (LCI) in 1997 to provide and share information about the management of construction projects in the most effective way. They observed that only 50% of the tasks on weekly work plans in construction projects are completed on time by foremen in a given week [13–15]. They proposed that construction practitioners can avoid these problems with active management of variability, starting with the structuring of the project (temporary production system) and continuing through its operation and improvement" [16]. This indicated that the construction industry is facing similar challenges to the manufacturing industry. Hence, the principles of the TPS and methods of Lean productions started to have been practiced in the industry by adapting them for construction.

Considering the similarity of challenges and need for improvement in both manufacturing and construction, the Lean methods have evolved with the methods for implementing. Hence, the main purpose of this chapter is to provide the background of Lean thinking in both manufacturing and construction along with presenting a bunch of Lean methods, which are widely practiced by industry practitioners. The chapter also mentions how Lean methods in production have changed when they are being implemented in the construction industry.

2. Background of lean production and lean construction: Interaction in terms of tools, techniques, and methods

Due to the quick industrialization after the industrial revolution, the world has become a place, where natural resources are unconsciously consumed and environmental problems increase. All these negative conditions have caused the run out of natural resources, distortion in the ozone layer, decrease in biodiversity, increase in environmental contamination, and global warming. Therefore, the removal of all these problems and negative conditions is one of the most important challenges of today's world. This leads to a considerable increase in the number of studies regarding the prevention of environmental problems, conscious use of natural resources, and a cleaner and healthier environment to be inherited by the next generations. In this context, Lean is a newly emerging concept for the majority of industry encouraging the effective use of resources. One of the major challenges of today's world is to execute projects more efficiently with respect to project objectives. At this point, Lean thinking aims to minimize waste while maximizing value to the customer.

Lean Production was the term coined by [17] to refer to Toyota's offering of high value, low-volume, and cost-competitive production to best address customer desires [18]. After the success of lean production in the automotive industry [19], Toyota's Lean thinking was applied in other industries. The construction industry produces more waste than any other industry in the entire world [20]. The waste oftentimes occurs in the form of workforce loss, safety breaches, material waste, and low efficiency. To avoid these, Lean construction has proven to be an effective means of production management for project delivery, i.e., designing and building capital facilities. Lean Construction is important in that it adopts the principle of minimizing waste and maximizing value while improving the total project performance per customer expectations. The need behind Lean construction comes from the failure of mass production and the persistence of craft-based production in the construction industry. Due to the changing needs of the customer, Lean construction is essential to provide the desired variety. To minimize waste and maximize value, researchers have previously focused on several different Lean construction methods. For example, it was implied that modular construction is effective in reducing waste and achieving resource efficiency [21]. This study also demonstrated that modular is reusable, which evidences the essential function of modular construction. In another study, it was indicated that there are several waste factors in mid/high-rise building projects and the determination of those waste factors is essential [22]. Therefore, Lean construction has proposed an opportunity for estimating the impacts of waste on overall project performance [23].

Sacks et al. [24] implied the importance of Lean production management systems in reducing waste in construction. Kalsaas [25] highlighted that measurement of waste and workflow is essential for the achievement of continuous improvement in construction projects. El.Reifi et al. [26] emphasized that Lean thinking is essential in the briefing process, where the design team develops their designs with respect to clients' desires. Fullalove [27] provided that the use of Lean techniques resulted in significant benefits such as an increase in return on investment and efficiency savings in UK road constructions. Marhani et al. [28] indicated that the application of Lean thinking into the construction industry provides a tremendous opportunity for the reduction of waste and an increase in production. Zhao and Chua [29] demonstrated that the reduction of non-value adding activities has a significant contribution to the construction productivity improvement. Aziz and Hafez [30] concluded that lean projects are safer, easier to manage, completed sooner, cost-effective, and are of better quality by referring to the impact of lean in minimizing waste in construction. Boyce [31] investigated the aspects of Lean thinking and concluded that it helps to improve the design phase of complex projects by emphasizing the essential function of a collaborative planning process in highway design. Going Lean is needed for the defective processes in mass production and craft production. Hence, Lean is an effective approach for customer satisfaction and enhanced project performance as previously implied by several studies [32, 33]. However, there is still a need for more effective Lean techniques to be applied in the construction projects especially given that the industry generally is reluctant to embrace and slow to adopt change.

Given this background, this chapter presents the most applied methods of Lean in the construction industry with inference to Lean production. The construction industry is utilizing most of the Lean techniques developed for manufacturing. Hence, it is essential to present these tools and techniques to guide industry practitioners for the proper implementation of the methods.

3. Lean methods: how tools and techniques are evolved

Lean methods have been heavily implemented in the manufacturing industry. Over time, the efficiency and reliability of the methods have been proven. This encouraged other industries to benefit from Lean methods. Since the construction industry relies on a heavy workforce, it is essential to utilize safer, reliable, and efficient methods and technologies.

In production, it is of utmost importance to eliminate 'waste'. Waste or 'muda' in Japanese is simply defined as anything other than the minimum amount of parts, materials, equipment, and work time specific to production [34]. There are seven waste types defined as overproduction, waiting time, transportation, inventory, processing, motion, and product defects. Lean manufacturing aims to manage processes without waste. However, it was evidenced that several companies are still challenging with staying Lean [35]. Kongguo [36] implied that Lean thinking helps conceive the Lean principles better, which first starts with realizing the customer value and continues with identifying value-added activities, generating flow, implementing the pull system, and sustaining continuous improvement. To improve the efficiency in those, various Lean methods and techniques are developed and practiced in manufacturing organizations. Some of them have been more effective in other industries such as construction.

Below are the widely implemented Lean techniques that have evolved and be used in the construction industry.

3.1 The last planner system (LPS)

LPS was originally developed by Glenn Ballard in 1993 in accordance with Lean construction principles. LPS is a Lean construction tool that focuses on increasing productivity by creating weekly work plans. The weekly plan includes tasks related to work and the individuals executing these tasks are called the Last planners [13]. LPS allows quick monitoring of the work-related issues for all construction personnel. LPS also provides an environment, where mistakes are visible. However, problems might occur, and timely actions are not taken in traditional construction management leading to late delivery of projects [36]. The Last planner is the person, who directly supervises the work. This person is usually responsible for production capability. The Last planner can be anybody like a project engineer, department manager, or foreman [37]. **Figure 1** presents the Last Planner System.

The tasks are split into two as needed and weekly. As needed tasks involve 'should' tasks, whereas weekly tasks include 'can', will', 'did' tasks. In 'should', the tasks include work to be done to reach the determined milestones according to the project plans. These tasks are created from different data such as customer demands, project goals, and information, planner stuff's former experiences. In 'can', the fundamental tasks are reflecting the actual work that is executed with respect to the constraints of the project. In this process, the required materials and labor are ready, where the previous project stage is completed. In 'will', the tasks ensure the work to be completed after all constraints are assessed. In 'did', the tasks refer to completed work [39].

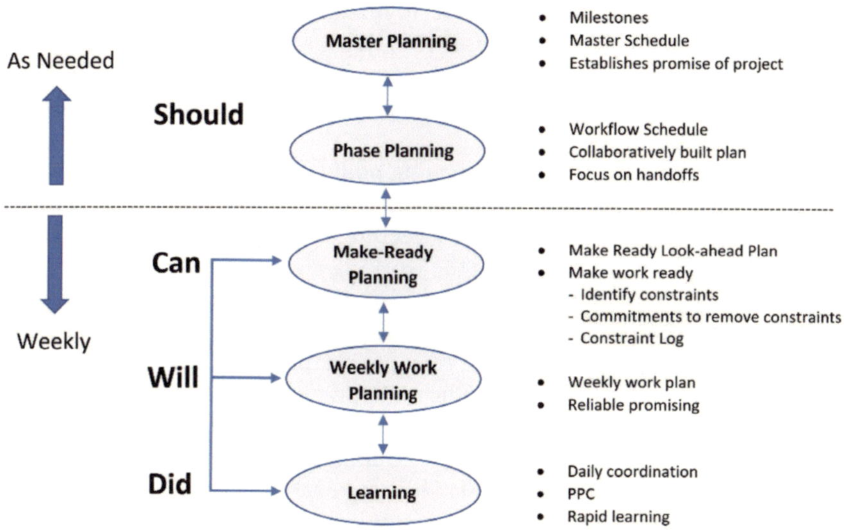

Figure 1.
Last planner system (adapted from [38]).

3.2 5S method

5S is a Japanese method of organizing the workspace in a clean, efficient, and safe manner to create a productive work environment. The 5S is a starting point for any company aiming to be recognized as a responsible and reliable producer [40]. In Japanese, the 5S methodology represents 5 different words, which all start with the letter S. **Figure 2** presents these five steps, respectively.

Sort (Seiri): Sorting is the first stage of 5S. It is the process of sorting out (separating) materials and equipment needed or unneeded. This process might result in fewer complaints, improved communication among employees, and an increase in the quality and efficiency of production. This process allows workers to take the next steps such as tagging the items.

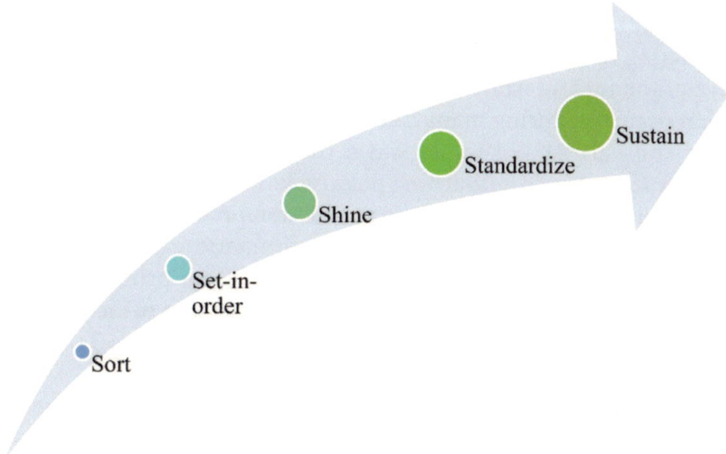

Figure 2.
5S stages.

Set in order (seiton): This stage refers to make all equipment needed for production accessible and prepared for use. This step also refers to organize all equipment and material for easy access and facilitation for production. This step requires the work area to be organized for production. A map can be drawn to represent station and equipment places.

Shine (seiso): This step refers to cleaning polluted equipment and work area. Pollution can be detected by sense organs and this might help find out the problem before it occurs. This stage also refers to sweeping everywhere cleanly and taking all kinds of unwanted objects away from the working environment. Thus, abnormalities can be noticed immediately, and the decision to clean materials after separation becomes easier.

Standardize (seiketsu): This stage refers to cleaning and maintaining the arrangement and standardizing that. The main purpose of this step is to fully meet 3S requirements and to detect and eliminate the root cause of problems. The way to ensure these is to constantly check the environment and detect deficiencies.

Sustain/self-discipline (shitsuke): This step encompasses all stages. It includes checking the existing system, training the employees, establishing good communication, and rewarding. The main purpose of this step is to get into the habit of maintaining the correct procedures [41].

3.3 Mistakeproofing (Poka yoke)

"Mistake proofing, or its Japanese equivalent poka-yoke (pronounced PO-ka yo-KAY), is the use of any automatic device or method that either makes it impossible for an error to occur or makes the error immediately obvious once it has occurred" [42]. Mistake proofing is an effective quality control technique to avoid human error, which might cause mistakes or defects [43]. Shingo [44] defines three inspection techniques for quality control, namely the judgment inspection, informative inspection, and source inspection. Judgment inspection is for discovering defects, whereas informative inspection is used to lower defect rates by controlling the process and prevent defects. Source inspection rather searches the conditions that exist for an error-free action.

Poka yokes might be grouped into three as shutdown poka-yoke, control poka-yoke, and warning poka-yoke in terms of their functions. The poka-yoke devices check different and important parameters and detect whether the process has an improper action. This check allows detecting whether the product manufactured has defects or not. The shutdown of poka-yokes constitutes an important part to prevent defects eliminating the possibility of error. The control poka-yoke is built into the production equipment and works as a redactor. When the device finds an unwanted condition that occurred during manufacturing processes, it signals production to avoid defects. The warning poka-yoke warns the operator with either visual symbols or sound signals for errors. The warning poka-yokes rely on human factors, where it is not quite certain to avoid defects in the production processes [45].

Mistake-proofing has six principles namely elimination, prevention, replacement, facilitation, detection, and mitigation. The first four principles intend to prevent the occurrence of human error, whereas the last two principles are to minimize the effects after the occurrence of human error. **Figure 3** presents these six principles along with their tasks.

The use of mistake-proofing devices also provides various advantages in terms of safety at the workplace [46]. It is possible to create fail-safe approaches in manufacturing with the use of such tools and devices. Considering the high accident rates in the construction industry, the use of mistake-proofing devices is also effective means of enhancing safety performance and avoiding human errors leading to work-related accidents.

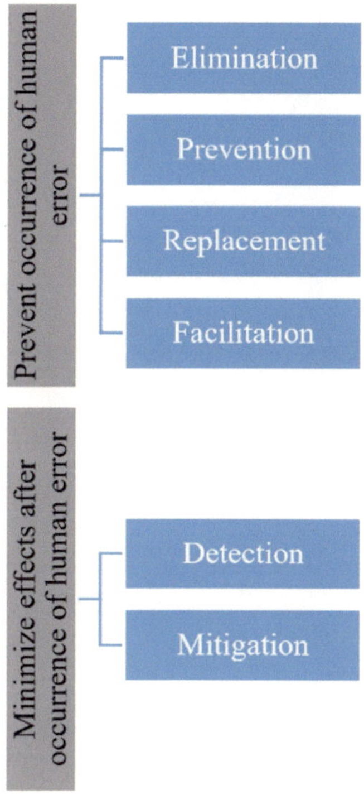

Figure 3.
Mistake proofing principles.

3.4 Visual management

Visual management is a broadly implemented Lean technique in the manufacturing industry. This technique helps to make information visible for all showing the information through visual signals [47]. Visual management has recently been used as a system enabling employees to better understand their role and contribution with respect to organizational values and customer needs. Nevertheless, the critical role of visual management has not yet been understood well by the construction industry. For example, two types of visual means such as 3D and visual planning are utilized in construction design [48]. Visual management helps increase communication, transparency, and stakeholders' capabilities [49, 50]. Therefore, construction companies must make use of these techniques to provide a better environment for their employees increasing efficiency and productivity.

3.5 Target value design (TVD)

Target Value Design (TVD) is simply defined as "a management practice that steers the design and construction of the project to the customer's constraints while maximizing the value delivered within those constraints" [51]. TVD is an emerging practice in the U.S. construction industry for cost predictability during design, construction, and delivery. It is adapted from the Target Costing method of manufacturing, which first appeared as a profit planning and strategic management approach in the 1930s [52]. This technique is promising for several benefits for the construction industry, where the companies are still struggling with project constraints such as cost, quality, and time. Therefore, TVD is an effective means of collaborative Lean approach in terms of reducing construction costs [53]. It was further indicated that

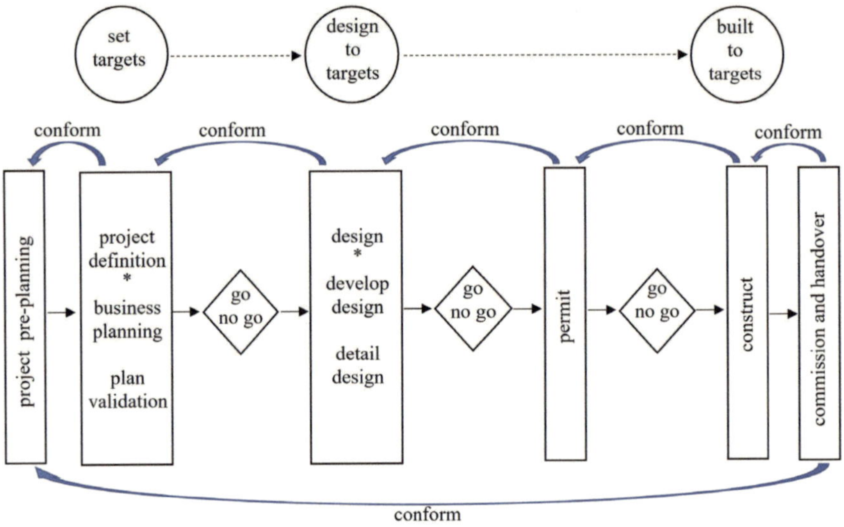

Figure 4.
TVD process scheme [53].

the systematic application of TVD resulted in significant improvement in project performance based on 12 construction projects, where TVD was introduced. **Figure 4** presents the TVD process with respect to construction project phases.

3.6 Value stream mapping (VSM)

Value Stream Mapping is an essential tool to identify and comprehend the productive stream focusing on the identification of waste sources, such as waiting for products and inventories, rework, information lost in the process, non-value-adding activities besides the identification of opportunities for improvement [54]. With VSM, it is possible to improve the information stream in the design process through the inclusion of alternative methods of control. This creates a base for incentives and future actions to generate value [55].

VSM helps visualize the whole rather than isolated parts of the process as well as monitoring the products, documents, and information. It also allows simultaneous visibility of streams of materials and information; visualization of indicators such as throughput time, percentage of value aggregation, lots size, and cycle time for the performance of activities [56].

VSM consists of several steps such as mapping activity for a family of products, defining the current state map of the value stream, and creating the future value stream map, where improvement takes place based on the proper identification of problems [54, 56]. **Figure 5** presents the steps for VSM.

3.7 5 whys and root cause analysis

5 Whys is a quality management tool of problem-solving aiming to find the root cause of an event [57]. It directs that one needs to ask five times repeatedly to identify the root cause of a problem for the fact that the solution is clear. This procedure aims to eliminate the root cause to prevent its recurrence [58]. **Figure 6** shows the 5 Whys procedure for finding the problem's root cause.

Considering the risky nature of construction projects, it is of utmost importance to determine the root cause of the problems leading to unwanted situations. Therefore, 5 Whys analysis is an essential method for preventing problems either

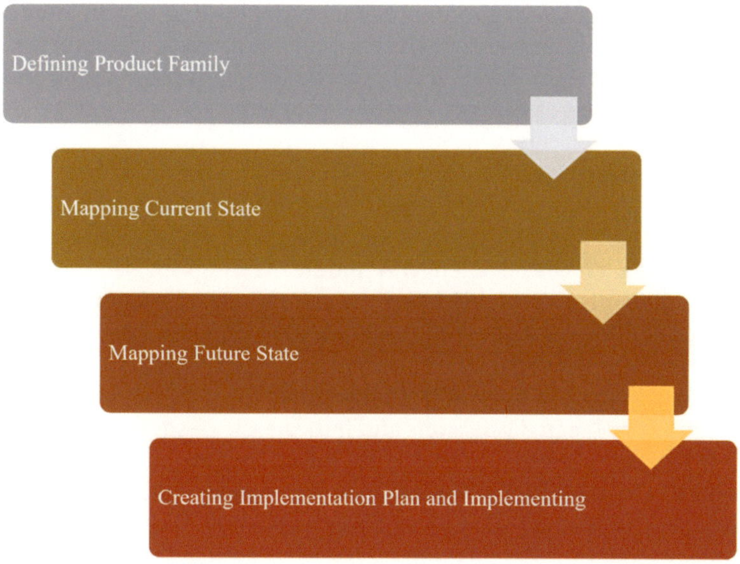

Figure 5.
VSM processes.

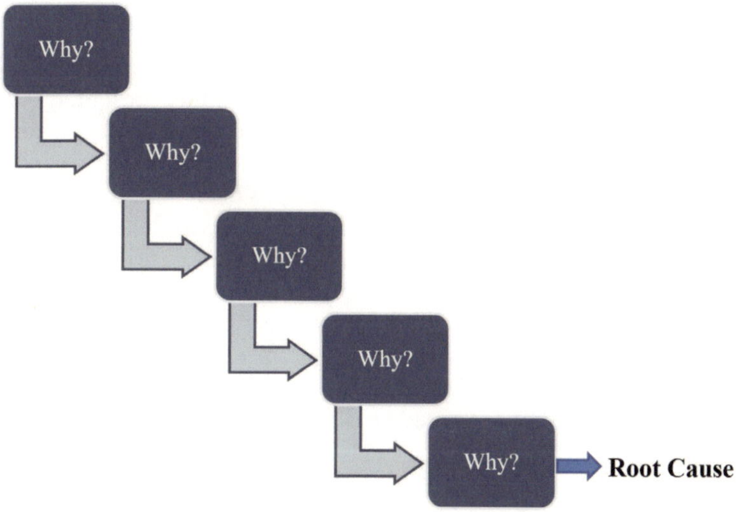

Figure 6.
5 whys analysis procedure.

from occurring or recurring. Therefore, utilizing the 5 Whys method might result in higher efficiency and productivity, where risky conditions are eliminated.

3.8 Gemba walks

Gemba is a Japanese word and it stands for the "actual place" [59]. For creating value in the organization, the actual place must enable employees to manufacture with less waste, fewer challenges, less overload, land ess overproduction. At this point, Gemba walks are essential to go and see the current situation and understand the root cause of the problem. In the Lean construction context, walking means "go see, ask why, show respect" [60]. Gemba walks help making the problems visible and create improvement ideas with the proper consideration of the root cause.

It also allows collecting data regarding the root cause leading to problems. In the construction industry, it is clear that Gemba walks constitute an important part since the majority of the processes in construction need improvements and require the proper identification of the root cause for problems.

3.9 Daily huddle meetings

Daily huddle meetings take place, where team members are ready to share what they achieved and what they challenge. A huddle meeting can also be organized as a weekly work plan meeting highlighting the completion of assignments for the following week in addition to discussing the work to be done that day [61]. The huddle meetings enhance the job satisfaction of employees while strengthening two-way communication among the team [62]. Daily huddle meetings create an opportunity for employees to involve in discussions and indicate the positive and negative sides of their tasks. The employees also find room for solving problems together during those meetings. These meetings also help detect the causes of accidents, which are associated with poor communication and coordination [63]. Hence, daily huddle meetings must be organized, and employees are encouraged to speak up on the tasks listing good and bad sides.

4. Conclusions

This chapter presented the historical evolution of Lean management and how Lean is adopted in the construction industry. The study presented the core principles of Lean along with the most widely adopted practices. According to the information presented in this chapter, one may advocate that the construction industry still struggling with the adaption of various Lean manufacturing practices into construction. Therefore, it is apparent that more research has to be conducted to provide a guideline for the industry practitioners in terms of benefitting from Lean practices at maximum. On the other hand, the methods, tools, and techniques presented in this chapter are expected to lead industry practitioners in terms of scrutinizing Lean concepts and evaluate those in the context of project conditions. As future work, the efficiency of Lean methods both applied in manufacturing and construction might be compared based on different operating processes.

Author details

Sevilay Demirkesen
Gebze Technical University, Kocaeli, Turkey

*Address all correspondence to: demirkesen@gtu.edu.tr

References

[1] Lean Enterprise Institute (LEI) 2021. A Brief History of Lean. https://www.lean.org/whatslean/history.cfm

[2] Melton T. The benefits of lean manufacturing: what lean thinking has to offer the process industries. Chemical engineering research and design, 2005, 83(6), 662-673.

[3] Womack, J. P., Jones, D. T., & Roos, D. (2007). The machine that changed the world: The story of lean production--Toyota's secret weapon in the global car wars that is now revolutionizing world industry. Simon and Schuster.

[4] Sorensen, Charles E., My Forty Years with Ford. New York: W.W. Norton, 1956.

[5] Schonberger, R.J. (2007) 'Japanese production management: an evolution – with mixed success', Journal of Operations Management, Vol. 25, No. 2, pp.403-419.

[6] Nordin, N., Deros, B. M., Wahab, D. A., & Rahman, M. N. A. (2012). A framework for organisational change management in lean manufacturing implementation. International Journal of Services and Operations Management, 12(1), 101. doi:10.1504/ijsom.2012.046676

[7] Womack, J.P., Jones, D.T. and Roos, D. (1990) The Machine that Changed the World: The triumph of Lean Production, Rawson Macmillan, New York.

[8] Womack, J.P. and Jones, D.T. (1996) Lean Thinking: Banish Waste and Create Wealth in your Corporation, Free Press, New York.

[9] Jaaron, A. and Backhouse, C.J. (2011) 'A methodology for the implementation of lean thinking in manufacturing support services', International Journal of Services and Operations Management, Vol. 9, No. 4, pp.389-410.

[10] Fairris, D. and Tohyama, H. (2002) 'Productive efficiency and the lean production system in Japan and the United States', Economic and Industrial Democracy, Vol. 23, No. 4, pp.529-554.

[11] Liker, J.K. and Hoseus, M. (2008) Toyota Culture: The Heart and Soul of the Toyota Way, McGraw-Hill, New York.

[12] Shapira, A., & Rosenfeld, Y. (2011). Achieving construction innovation through academia-industry cooperation—Keys to success. Journal of Professional Issues in Engineering Education & Practice, 137(4), 223-231.

[13] Ballard, G. and Howell, G. (1994a). "Implementing Lean Construction: Stabilizing Work Flow." Proceedings of the 2nd Annual Meeting of the International Group for Lean Construction, Santiago, Chile.

[14] Ballard, G. and Howell, G. (1994b). "Implementing Lean Construction: Improving Performance Behind the Shield." Proceedings of the 2nd Annual Meeting of the International Group for Lean Construction, Santiago, Chile.

[15] Ballard, G. and Howell, G. (1998). "Shielding Production: Essential Step in Production Control". Journal of Construction Engineering and Project Management, Vol. 124, No. 1, pp. 11-17.

[16] Ballard, G. and Howell, G. (2003). "Competing Construction Management Paradigms" (PDF). Proceedings of the 2003 ASCE Construction Research Congress. Honolulu, Hawaii. Retrieved 31 March 2013.

[17] Krafcik, J. F. (1988), "Triumph of the Lean Production system", Sloan Management Review, Volume 30, Issue 1,

[18] Lean Enterprise Institute (2021), http://www.lean.org/WhatsLean/History.cfm (accessed April 17, 2020).

[19] Womack, J. P., Jones, D. T., Roos, D., (1990), "The Machine that Changed the World", Harper Perennial, New York.

[20] Meadows, D., (2011), "Designing out waste. Environmental Design & Construction", http://www.modular.org/marketing/documents/DesigningoutWaste_EDC.pdf (accessed July 29, 2020).

[21] Alshayeb, M.J., (2011), "Lean Production Using Modular Construction", Master's Thesis, Engineering Management Program and the Faculty of the Graduate School, The University of Kansas, USA.

[22] Khanh, H.D., Kim, S.Y., (2014), "Evaluating impact of waste factors on project performance cost in Vietnam", KSCE Journal of Civil Engineering, Volume 18, Issue 7, pp 1923-1933

[23] Ali, S.A., Khadem, M., Seifoddini, H., (2008), "Efficacy of lean metrics in evaluating the performance of manufacturing systems". International Journal of Industrial Engineering, 15, pp.176-184.

[24] Sacks, R., Radosavljevic, M., Barak, R., (2010), "Principles for Building Information Modeling based Lean Production Management Systems for Construction", Automation in Construction, Vol. 19 No. 5 pp. 641-655.

[25] Kalsaas, B.T., (2013), "Measuring Waste and Workflow in Construction", 21th Annual Conference of the International Group for Lean Construction. Fortaleza, Brazil, 31-2 Aug 2013. pp 33-42

[26] El.Reifi, M.H., Emmitt, S., Ruikar, K., (2013), "Developing a Conceptual Lean Briefing Process Model for Lean Design Management", 21th Annual Conference of the International Group for Lean Construction. Fortaleza, Brazil, 31-2 Aug 2013. pp 329-338

[27] Fullalove, L.H., (2013), "Examples of Lean Techniques and Methodology Applied to Uk Road Schemes", 21th Annual Conference of the International Group for Lean Construction. Fortaleza, Brazil, 31-2 Aug 2013. pp 1057-1066.

[28] Marhani, M.A., Jaapar, A., Bari, N.A.A., Zawawi, M., (2013), "Sustainability through Lean Construction Approach: A Literature Review", Procedia - Social and Behavioral Sciences, 8 November 2013, Pages 90-99

[29] Zhao, Y., Chua, D.K.H., (2003), "Relationship between productivity and non value-adding activities", Proceedings of the 11th annual conference of the International Group for Lean Construction, Blacksburg, Virginia

[30] Aziz, R., Hafez, S. (2013). Applying Lean Thinking in Construction and Performance Improvement. Alexandria Engineering Journal, 52, 679-695.

[31] Boyce, E., Dainty, A., Thorpe, A. (2012). A Novel Collaborative Planning Methodology for Complex Infrastructure Design Projects. N. Thurairajah, ed. In: CIB Joint Symposium, Management of Construction: Research to Practice, Montreal, Canada, June 26-29.

[32] Horman, M., Kenley, R. (1996). The Application of Lean Production to Project Management. Proceedings of the 4th Meeting of the International Group for Lean Construction (IGLC), University of Birmingham, UK.

[33] Khadem, M., Ali, S., Seifoddini, H., (2008). Efficacy of lean metrics in evaluating the performance of

manufacturing systems. International Journal of Industrial Engineering, 15 (2), 176-184.

[34] Rahman, N. A. A., Sharif, S. M., & Esa, M. M. (2013). Lean manufacturing case study with Kanban system implementation. Procedia Economics and Finance, 7, 174-180.

[35] Shahram, T., 2007. Lean manufacturing performance in China: assessment of 65 manufacturing plants, Vol.19, No. 2, p. 217-234.

[36] Kongguo, Z. (2014, May). Research on the emergence mechanism of Last Planner System of lean construction. In The 26th Chinese Control and Decision Conference (2014 CCDC) (pp. 3643-3646). IEEE.

[37] Ballard, G., Hamzeh, F. R., & Tommelein, I. D. (2007). The Last Planner Production Workbook-Improving Reliability in Planning and Workflow. Lean Construction Institute, San Francisco, California, USA, 81.

[38] Ballard, G., & Tommelein, I. (2016). Current process benchmark for the last planner system. Lean Construction Journal, 89, 57-89.

[39] Gao, S., & Low, S. P. (2014). The Last Planner System in China's Construction Industry—A SWOT Analysis on Implementation. International Journal of Project Management, 32(7), 1260-1272.

[40] Veres, C., Marian, L., Moica, S., & Al-Akel, K. (2018). Case study concerning 5S method impact in an automotive company. Procedia Manufacturing, 22, 900-905.

[41] Patel, V. C., & Thakkar, H. (2014). Review on implementation of 5S in various organization. International Journal of Engineering Research and Applications, 4(3), 774-779.

[42] ASQ (2020). "What is Mistake Proofing?" https://asq.org/quality-resources/mistake-proofing

[43] Grout, J. R. (1997). Mistake-proofing production. Production and inventory management journal, 38(3), 33.

[44] Shingo, S. (1986). Zero quality control: source inspection and the poka-yoke system. CRC Press.

[45] Tommelein, I., Demirkesen, S. (2018). Mistakeproofing The Design of Construction Processes Using Inventive Problem Solving (TRIZ). CPWR-The Center for Construction Research and Training, CPWR Small Study Final Report, Silver Spring, MD

[46] Saurin, T. A., Formoso, C. T., & Cambraia, F. B. (2008). An analysis of construction safety best practices from a cognitive systems engineering perspective. Safety science, 46(8), 1169-1183.

[47] Singh, S., & Kumar, K. (2020). A study of lean construction and visual management tools through cluster analysis. Ain Shams Engineering Journal.

[48] Tjell, J., & Bosch-Sijtsema, P. M. (2015). Visual management in mid-sized construction design projects. Procedia Economics and Finance, 21, 193-200.

[49] Liker, J. K. (2004). The 14 principles of the Toyota way: an executive summary of the culture behind TPS. The Toyota Way, 14(1), 35-41.

[50] Tezel, B. A., Koskela, L. J., & Tzortzopoulos, P. (2010). Visual management in construction: Study report on Brazilian cases.

[51] De Melo, R. S. S., Do, D., Tillmann, P., Ballard, G., & Granja, A. D. (2016).

Target value design in the public sector: evidence from a hospital project in San Francisco, CA. Architectural Engineering and Design Management, 12(2), 125-137.

[52] Cooper, R., & Kaplan, R. (1999). Design of cost management systems. Upper Saddle River, NJ: Prentice Hall.

[53] Zimina, D., Ballard, G., & Pasquire, C. (2012). Target value design: using collaboration and a lean approach to reduce construction cost. Construction Management and Economics, 30(5), 383-398.

[54] Rother, M., & Shook, J. (1999). Learning to See: Value Stream Mapping to Add Value and Eliminate Muda. The Lean Enterprise Institute. Inc., Brookline, MA.

[55] Freire, J., & Alarcón, L. F. (2002). Achieving lean design process: Improvement methodology. Journal of Construction Engineering and management, 128(3), 248-256.

[56] Womack, J. P. (2006). Value stream mapping. Manufacturing engineering, 136(5), 145-156.

[57] Ansah, R. H., Sorooshian, S., Mustafa, S. B., & Duvvuru, G. (2016, September). Lean construction tools. In Proceedings of the 2016 International Conference on Industrial Engineering and Operations Management Detroit, Michigan, USA.

[58] Tsao, C. C. Y., Tommelein, I. D., Swanlund, E. S. and Howell, G. A. 2004. Work Structuring to Achieve Integrated Product-Process Design. Journal of Construction Engineering and Management. Vol. 130, Iss. 6. pp. 180-189.

[59] Imai, M. (2007). Gemba Kaizen. A commonsense, low-cost approach to management. In Das Summa Summarum des Management (pp. 7-15). Gabler.

[60] Womack, J. (2011). Gemba Walks. Lean Enterprise Institute.

[61] Zhang, L., & Chen, X. (2016). Role of lean tools in supporting knowledge creation and performance in lean construction. Procedia Engineering, 145, 1267-1274.

[62] Ogunbiyi, O., Oladapo, A. and Goulding, J. (2013), "A review of lean concept and its application to sustainable construction in the UK", International Journal of Sustainable Construction Engineering and Technology, Vol. 4 No. 2, pp. 82-92.

[63] Enshassi, A., Saleh, N., & Mohamed, S. (2019). Application level of lean construction techniques in reducing accidents in construction projects. Journal of Financial Management of Property and Construction.

Chapter 7

Model-Based Enterprise Continuous Improvement

Bruno Vallespir and Anne Zouggar-Amrani

Abstract

The enterprise reengineering based on enterprise modelling is usually carried out within the framework of conventional projects. This leads to relatively long projects that are not compatible with a highly variable economic environment. The objective of the evolution management presented here is to use enterprise modelling and all the benefits it brings in a framework that allows for more continuous improvement than is generally observed. The proposed architecture is made up of three levels: a strategic level based on performance measurement, a tactical level that manages system migration and is based on enterprise models, and an operational level consisting of managing a portfolio of evolution projects. Together, these allow a shorter set of projects to be carried out, while remaining coherent and aligned with the company's strategy. This approach puts enterprise modelling methods and continuous improvement/Lean management approaches into perspective, allowing complementarities and opening up interesting perspectives concerning enterprise re-engineering methods.

Keywords: enterprise modelling, evolution management, continuous improvement, lean management, performance

1. Introduction

Since the 1970s, enterprise modelling has developed into an effective methodological source for improving business performance. Some of the proposed approaches simply provide a modelling language but others also present an implementation method. It appears that these methods adopt a classic project approach that leads to long and costly projects. Moreover, in the context of a rapidly changing economic environment, these approaches lack responsiveness. Faced with this, continuous improvement is pushing towards shorter projects that come from the field and are part of a permanent movement of evolution.

With this perspective in mind, the objective of this chapter is to show how enterprise modelling can be encapsulated in a continuous evolution approach of a strategic nature, the ultimate goal being to take advantage of the expressiveness and systemic approach of enterprise modelling while being part of a fluid and reactive evolution context.

The outline of this chapter is as follows. Section 2 will present the problem statement by insisting on the inadequacy of project-based approaches in a context of a changing environment. Section 3 gives elements of conceptualisation, on the one hand, on the evolving system itself and, on the other hand, on the system for managing this evolution. Section 4 will present the evolution management system in detail. Section 5 will give the main elements that argue in favour of such an approach. Section 6 will conclude the chapter.

2. Problem statement

Over the last few decades, enterprise modelling has provided a methodological set of tools for engineering and, more often, re-engineering organisations. Little by little, this scientific field has emerged as an effective methodological source for improving business performance [1–3]. Developments took place in several stages [4, 5]. After having proposed many modelling languages in the 70s and 80s, this field then sought to make these languages work together to obtain integrated methods (such as CIMOSA or GIM) with a large modelling coverage in order to approach companies in the most systemic way possible [6–8]. This work made it possible to define fairly stable modelling domains, often identified as views or points of view: informational view, process view, decisional view, etc. The next step consisted in organising all this input by analysing on the one hand the components of these methods and their organisation (GERAM) [9, 10] and on the other hand on the nature of the concepts handled. This last point was based on approaches such as meta-modelling and ontologies and had as a practical field of application the translation of inter-language models and the development of a Unified Enterprise Modelling Language (UEML) [11–13]. From a theoretical point of view, this point allowed the identification of the major concepts to be retained in enterprise modelling as well as the way to formalise and express them. Finally, it must be stressed that enterprise modelling corresponds well to current trends that advocate the use of models in engineering such as Model-Driven Architecture (MDA) [14, 15] in software engineering or all the approaches referenced under the term Model-Based Systems Engineering (MBSE) [16].

Applications of enterprise modelling methods show that they lend themselves well to project-based approaches. Project management in companies has grasped big attention since many decades to provide new insights to the practitioners. Early investigation through case studies in [17] provides a cross analysis between project management and the interest of Lean thinking. A key element in combining lean approach to project is "Planning and control by objectives" with fixed and accepted key dates. Then, the commitment and motivation from the team was quoted as leading to successful final project. This link requires precise organisation and timing, time and resources. A complete project of this type takes place over several months and can take up to one year.

It is emphasised in [18] that the efficient resources management is becoming a major challenge in the current context of volatility, uncertainty, complexity and ambiguity. In order to efficiently manage its resources, companies need to manage and deliver projects on time, on budget, inside the scope and in accordance with the quality requirements agreed with the customer. We are therefore faced with the two classic problems of this type of approach.

The first problem concerns the evolution of the environment, and therefore of the specifications, during the project. Like any project, a reengineering project using enterprise modelling is based on initial specifications and objectives. Even if it is possible to make these evolve during the project, it is more comfortable and efficient to ensure that they remain fixed for the duration of the project. In the end, a project-based approach is easier to implement in a stable context ensuring that the specifications do not change significantly during the project.

The second problem concerns the necessary breaks between projects. These are necessary for several reasons. Firstly, a re-engineering project is sufficiently intrusive and impacting that the system under consideration needs to "rest" between projects i.e., to return to a nominal regime during which the project results will be integrated into the day-to-day running of the company. Secondly, as this type of project requires a financial and time investment, this effort cannot last indefinitely. The break thus enables the company to reconstitute its resources before considering another project. Generally speaking, it can be envisaged that the return on investment must be sufficient before considering another project. In the end, since the break is necessary, the project will be all the more profitable if requirements do not change too quickly during the break. This brings back to the necessary stability of the environment.

In conclusion, the major problem is the stability of the environment. The project approach is difficult to apply in a turbulent context. **Figure 1** summarises these points.

The answer to this problem is therefore to reduce these two durations: project duration and the duration of break between projects. The solution is to move towards less ambitious and more targeted projects, even if it means multiplying them. A less ambitious project can be carried out more quickly. Because it is shorter, there is less risk of a gap between specifications and results. A less ambitious project also requires fewer resources, which makes it easier to make it profitable. Finally, a less ambitious project has less impact on the entire structure, which makes it easier to integrate the results in nominal mode. These last two points thus limit the need for break between projects. **Figure 2** shows how shorter but more numerous projects, with shorter breaks between projects, can make it easier to meet the company's expectations.

This orientation leads to a more continuous evolution of the system. Therefore, we are approaching methods referred to as continuous improvement. In [19] it is reminded that project management model suggests to systematically "address the actions and solutions to be implemented in order to keep, in the long run, the continuous improvement of the project management processes in the organization".

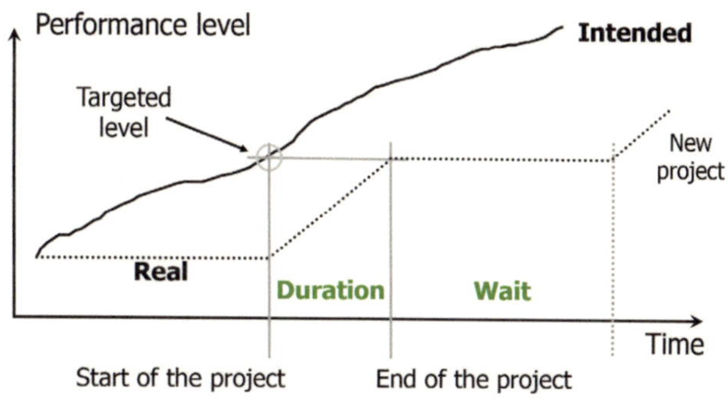

Figure 1.
The problems issued from a project-based approach in a turbulent environment.

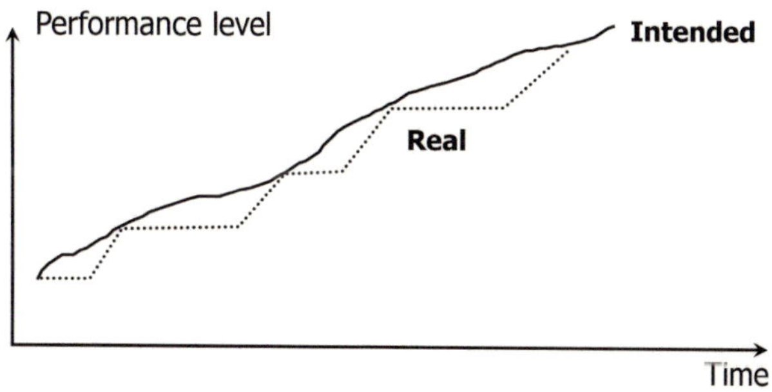

Figure 2.
Getting closer to the needs through shorter, more numerous projects and with less break between projects.

The DMAIC (define, measure, analyse, improve, control) is also sustained as being a cycle for conjoint continuous improvement framework [20]. The DMAIC methodology is seen as last generation of improvement approaches, adding concepts, methods, tools and removing limitations identified [21]. The model based on DMAIC allowed identifying company's main project management problems and associated causes and the selection of the causes to be first addressed [19]. It is closely linked to PDCA approach evoked further.

This field, which has a very strong intersection with Lean management [22, 23], proposes a philosophy and a set of methods that provide tools for improvement actions. The Lean thinking is a way of focusing on value from customer point of view and making people contributing to the improvement to ensure the quality at the source. When the actions carried out with Lean practices such as Value Stream Mapping, Kaizen, A3 approach are examined, it effectively shows that they are less ambitious and more focused on a specific problem. Starting from problems in the field and involving various company members, they generally focus on the physical system (in the industrial case) or, more generally, on the value-added process to get as much exhaustive vision of the flow as possible and to analyse operational dysfunctions. The analyses of the added value activities should and must be at the heart of the focus that leads to less interest in infrastructural items such as the information system.

In addition, they offer more problem-solving tools than enterprise modelling. Conversely, this results in a weaker systemic vision than with enterprise modelling (how do all these actions fit into a coherent whole?). Similarly, it presents very few representation tools unlike enterprise modelling. Only Value Stream Mapping (VSM) can be considered as a modelling language. As quoted in [24], VSM is a powerful tool of representation found as being able to eliminate Muda, bottlenecks across production line. The value stream mapping uses current state map to record current state of production line before implementation of improvements. Indeed, the VSM contains a specific pictograms code to represent steps of the flow along the considered scope (from suppliers to customers) with different technical data at each activity represented. The information and physical flows are modelled to visualise the flow progression and detect "bottleneck resources" that deserves attention and corrective actions. By the way, VSM modelling is also significantly interesting tool to perceive the durations of the added value actions and the waste undergone in the different steps because of storages, quality rate and processing times. VSM was efficiently proved to be interesting in the modelling production flow of an aeronautic company to improve the productivity and deliveries costs dropped by 50% [25].

Generally speaking, what most characterises continuous improvement is the continuous aspect of the actions carried out, as the name suggests. Here, there is no project with a beginning and an end, but a continuous improvement process, conceptualised in particular by the Plan-Do-Check-Act cycle (PDCA) which is a control framework for executing a series of activities for continuous improvement of processes, originally developed in the field of manufacturing [26].

Finally, the approach presented in this chapter aims to move towards an approach of continuous evolution of the system under consideration, while retaining the advantages of modelling as proposed by enterprise modelling. To avoid confusion with continuous improvement, the approach is referred to here as evolution management.

3. Conceptualisation

Several aspects concerning conceptualization are presented in this part. Firstly, the notion of evolution trajectory makes it possible to implement the conclusions of the previous part. Then, several levels of management are proposed to manage the evolution trajectory of the system. Finally, several ways of formalising the system are presented [27–29].

3.1 Evolution trajectories

The general principle is then to make the evolution of the system a process as continuous as possible. Practically, the evolution process is made up of a sequence of steps representing the evolution of the state of the system. The closer in time these steps are, the more continuous the evolution of the system will be. Two steps are specific. The first one corresponds to the state of the system at the time it is examined (t = 0). This step therefore corresponds to the *current state*. The second represents the state in which we would like the system to be in the future, at a time sufficiently far in the future but for which it is possible to make viable predictions about the system's environment. We refer to this step as the *target* and the moment at which it corresponds as the *strategic horizon*. The path between the current state and the target is punctuated by intermediate states that we call *steps*. These steps are the moments when the environment is reassessed and the target is redefined. If the environment has not changed, the target remains the same. This is equivalent to saying that the target is the desired state in the future, assuming the environment has not changed. However, we will consider that this is not the general case. Therefore, at each step, a new target is defined. The duration between two steps is usually fixed, we call this duration *strategic period*. It is clear that, because the steps are intended to be moments of redefinition of the target, the strategic period will be all the shorter as the environment changes rapidly.

Figure 3 summarises these concepts.

3.2 Management levels

On the basis of the trajectory of the system as we have just defined it, several levels of management can be envisaged.

The first one corresponds to the control of the path between the current state and the target. The target is a state envisaged at long term, based on the analysis of the environment and the company's major orientations with a significant degree of uncertainty. The concept of target is close to other concepts such as vision, mission or values which are the core elements of a strategic organisational foundation [19]. Therefore, it corresponds to a strategic level.

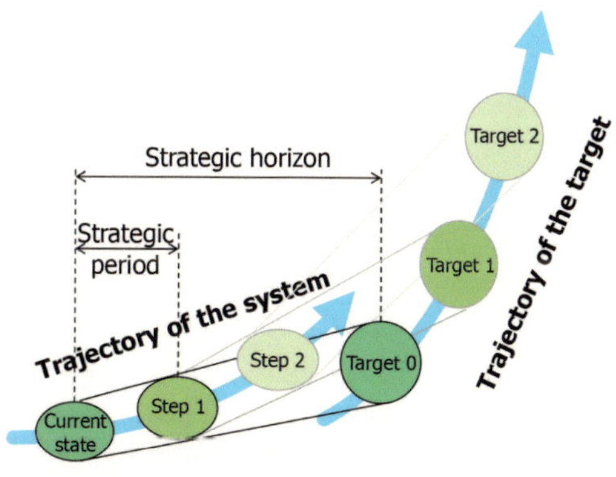

Figure 3.
Evolution trajectories of the system and the target.

As much as the target is considered to be generally unreachable, step 1 must be reached (there is no questioning planned before step 1). The management of the evolution between the current state and step 1 must therefore make it possible to precisely define the state of the system at step 1. This level is therefore considered tactical.

The level that has just been presented makes it possible to define towards which state the system must evolve, but it does not manage the actions to be implemented to do so. Therefore, a third level, concerned by concrete action, is necessary. This level is operational.

Figure 4 shows these three levels and the processes that they manage. Considering the role that they play in the approach, the current state is called *As-is*, the first step *To-be* and the target *Could-be*.

3.3 Formalisation modes

The states identified by the approach can be formalised in different ways. Three forms are envisaged: performance, model and project.

Performance. A system can only be seen as a source of performance. Once the set of performances of interest to the company has been defined, the system and its evolution will be characterised through these performances. The state of the system can therefore be considered to change each time a performance changes in value. Thus, the state of the system is characterised by the value of its performance vector. The evolution then becomes a trajectory in a performance space, the significant points of this evolution being the states of interest. The performance can be observed in the case of an existing system or targeted in the case of a future system. **Figure 5** illustrates this approach.

Model. The most classic way to represent a system is to make a model of it. The notion of model is very broad and the definition of this term changes according to the domains. In engineering, a model represents the structure or behaviour of a system and is intended either to understand and evaluate the system when it exists

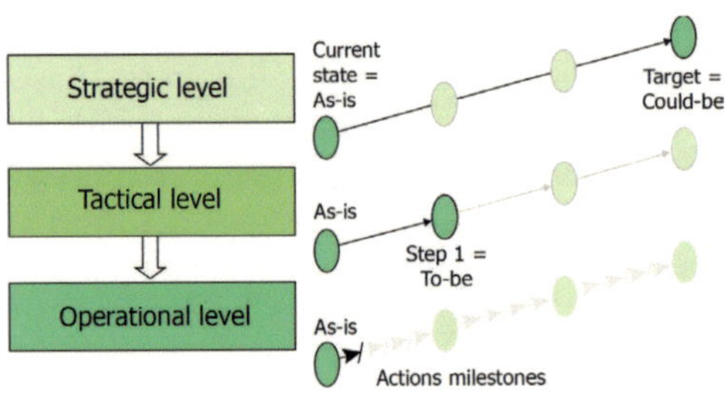

Figure 4.
The management levels.

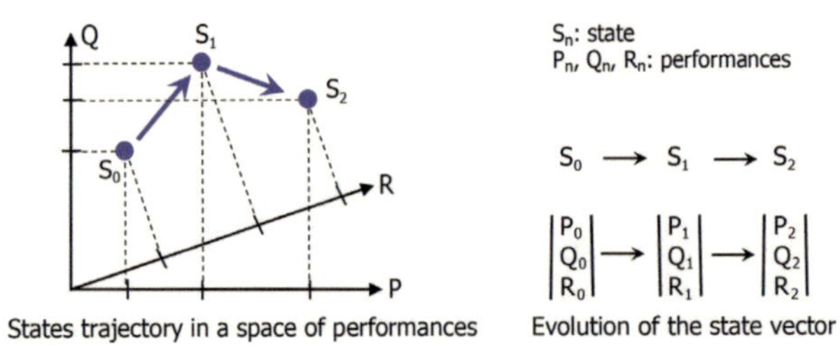

Figure 5.
The performance-based characterisation.

or to characterise it in order to design it. A model is based on a language. It can be formal, semi-formal or informal. A formal language is based on a mathematical formulation, whether continuous (system of differential equations for example) or logical (discrete event systems for example). At the other end of the spectrum (informal models), we can find models that are only drawings. A shop layout is an example of this. In between are semi-formal languages i.e., languages that have syntax and lead to less interpretation than natural languages but are not executable. This is the domain of enterprise modelling. The latter proposes a set of approaches and graphical languages that allow the system to be observed from several points of view. These languages include the IDEF suite, the business process modelling languages (BPMN, ...), the GRAI method, CIMOSA, etc. The aim here is not to define the language to be used, this depends on the objectives of the company and its culture. Finally, we should not forget simulation, which is quite similar to enterprise modelling but which proposes executable models.

Project. A final way of understanding the system is through the projects it undergoes. This way, less classical than the two previous ones, insists on the fact that an evolving system is the object of projects that act on it and that, therefore, the evolution of the system is characterised by the projects that allow it. Within this framework, future projects can be envisaged to support a targeted evolution and current projects can be analysed to understand the evolution in progress. Finally, looking at the projects means observing the evolution of the system in an operational way.

The three approaches are complementary. Seeing the system through its performances consists in considering it as a black box and in valuing the exchanges it implements. The model approach allows on the one hand to open the black box to observe the structure and, on the other hand, to observe the dynamics of the system (synchrony). Finally, the vision by project focuses on a diachronic approach by analysing the actions that lead the system to evolve.

4. The evolution management system

The general architecture of the evolution management system is based on the elements of conceptualisation presented by the previous chapter. It is structured on three levels.

The first level, entitled "*Strategic orientation*", is intended to propose a path leading from the current state (as-is) to the target (could-be) over the strategic horizon. This path is made up of regular steps. The strategic orientation level is expressed in terms of performances for two reasons. Firstly, given its nature, it makes it easier to link it to the strategy of the company. Secondly, because the target and all the steps following the first one will not be reached a priori, it saves an unnecessary effort of formalisation. The result of this level is a level of performances for each step.

The second level is called "*Migration plan*". Its objective is to express the path from the current state (as-is) to the first step (to-be) over the tactical horizon (that is equal to the strategic period – **Figure 3**). Knowing that this step must be reached, a modelling action deserves to be carried out. Therefore, this level works on the basis of models. This level leads to the definition of the models of the first step and of the set of actions to be implemented to reach it.

The third level is called "*Projects portfolio*". On the basis of the migration plan defined at the level above, the objective of this level is to define the projects operationally and to ensure the management of the entire projects' portfolio (over the tactical horizon) and all projects individually (over the project duration).

Table 1 shows the overall picture.

Name of the level	Nature	Expression mode	Initial state	Final state	Horizon
Strategic orientation	Strategic	Performance	Current state (As-is)	Target (Could-be)	Strategic horizon
Migration plan	Tactical	Model	Current state (As-is)	First step (To-be)	Tactical horizon (Strategic period)
Projects portfolio	Operational	Project	Current state (As-is) / Project start	Portfolio completion / Project end	Tactical horizon / Project duration (Operational horizon)

Table 1.
Architecture and principles of the evolution management system.

4.1 The strategic orientation level

As already explained (**Table 1**), there is several time milestones organising this level. Upstream, there is the existing state, corresponding to the system as it is now (as-is). Downstream, there is the target that is the representation of the system as we would like it to be at the strategic horizon, assuming that no significant element of the environment would change between now and then (could-be). The target is therefore positioned at the furthest point in the future at which it is possible to make assumptions about the system. In between, steps are distributed at regular intervals (strategic period). In theory, the number of steps is equal to the strategic horizon divided by the strategic period. The steps correspond to the moments when the trajectory to be followed is questioned.

All these milestones express the system in terms of performances. As explained above, that means that the system is positioned in a performances space.

The three main activities implemented at this level are as follows.

1. *Target definition*. This consists in translating the "key success factors" provided by the company's strategy into a valued technical performances vector. The nature of these performances is decided by the company itself. There is a double condition about these performances: in one hand, to be valuable on the basis of key success factors and, in another hand, to be operational enough to support the definition of change about the system.

2. *Current state evaluation*. This action consists of evaluating the existing situation in the same performance vector as for the target. As we are dealing here with the existing situation, this evaluation can be carried out on the basis of observations and measures. In comparison with the target definition, the distance in terms of performances can be calculated.

3. *Trajectory definition* (steps). On the basis of the distance value calculated in the previous action, the objective of this action is to define a steps trajectory between current state and target, knowing that there must be one step for each strategic period. The steps are expressed with the same performances vector.

Figure 6 summarises these activities.

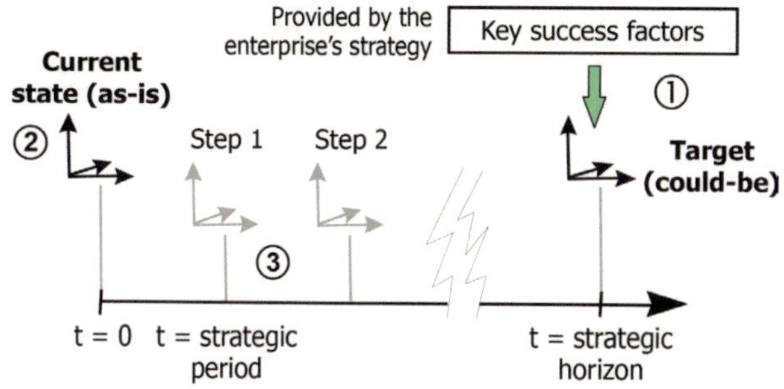

Figure 6.
The activities of the strategic orientation level.

The definition of the trajectory and, therefore, of the steps that constitute it, is not a simple task for two main reasons. Firstly, it can be difficult to translate the key success factors, often expressed in general terms, into operational objectives i.e., objectives that are valued and translatable into actions. Secondly, the performance space is not accessible in its entirety. The reason for this could be:

- the exogeneous limitation of the level of a performance (technical, legal, etc.);
- the deadly cost of making a certain level of performance accessible;
- the fact that some performances may be opposite: seeking to increase one inevitably leads to reducing the level of another;
- the fact that some performances may rely on the same type of resources (financial or other) that are inevitably limited, this leads to finding a compromise in the distribution of this resource between the two performances.

4.2 The migration plan level

The two time milestones structuring this level are the current state and the first step. These two milestones have already been explained and are present at the level above (**Figure 6**). The difference with the previous level is that here they are expressed in the form of models. This transition, from an expression in terms of performances to a representation by models, corresponds to an operationalisation process i.e., a willingness to move towards a concrete vision. This is justified at the level of the first step since this will be reached and therefore corresponds to an implemented state.

It is not the purpose of this chapter to propose one enterprise modelling approach over another. There are many business modelling methods and languages available and the choice will have to be made according to the culture of the company. It is always important to cover all the views considered important in a modelling approach: processes, data, physical system, decisions, organisation, etc. To do this, it will be possible to choose languages each corresponding to one of these views or to use multi-point of view methods that already integrate several languages (GIM or CIMOSA, for example). In any case, we consider that the approach proposed here works independently of the languages chosen.

The three main activities structuring this level are as follows.

1. *Current state modelling*. This action consists of modelling the system in its existing situation, in terms of structure and behaviour. This action concerns

the current state. Then, it is possible to use the whole set of instruments available to an analyst to build the model of an existing system: consultation of documents, analysis of computer application screens, field observations, interviews, etc. This action must be able to propose, in complement to the models themselves, an analysis of the system in terms of strengths and weaknesses.

2. *First step modelling*. This action consists in proposing a model of the system which, as a priority, allows to translate the level of performances defined for the first step by the upper level. This model must also take into account the shortcomings observed in the current state of the system (as-is model) and the possible evolution needs expressed by the company. In addition, the model will need to preserve the strengths identified in the previous action. Here we are in a totally different situation compared to the previous action. The modelling of the existing state was based on observation, the modelling of a future state is based on creativity. We are therefore here at the heart of the engineer's job, which is to propose the model of a future system, based on the expression of needs and expected performances, with all the uncertainty that it entails.

3. *Actions plan*. The evaluation of the difference between the model of the first step and the model of the current state enables the definition of a list of actions necessary to evolve i.e., to make the system moving from its current state to the first step. The aim here is not to carry out these actions but to define them, taking into account the fact that they are interdependent. Because of this interdependence (an action needs that another one must be proceeded before, for instance) and because the resources of the company are obviously limited, these actions must be sorted in terms of priority.

Figure 7 represents these activities.

We are here in the typical enterprise modelling context: an instance of the migration plan corresponds to an enterprise modelling project. Obviously, the objective here being to converge towards a continuous evolution, the migration between the existing state and the first step will thus correspond to a less ambitious evolution than what classically constitutes the perimeter of a project. Nevertheless, the principle remains the same. To illustrate this, **Figure 8** presents the general principle of conceptualisation followed by enterprise modelling [30], also known as the *"sun curve"* in information systems design (1. *modelling*: passage from the reality of the existing state to its model, 2. *analysis and design*: passage from the model of the

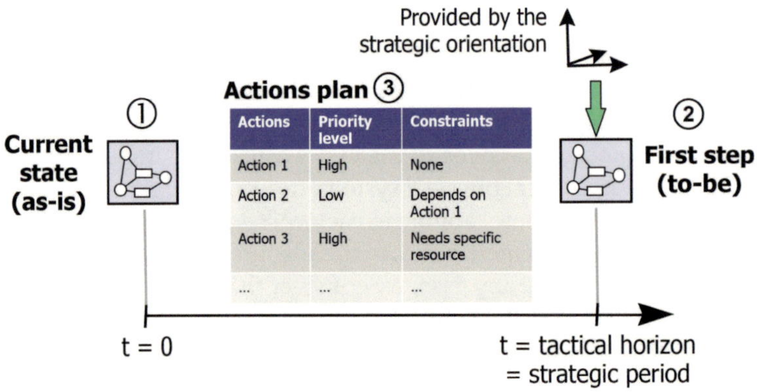

Actions	Priority level	Constraints
Action 1	High	None
Action 2	Low	Depends on Action 1
Action 3	High	Needs specific resource
...	...	...

Figure 7.
The activities of the migration plan level.

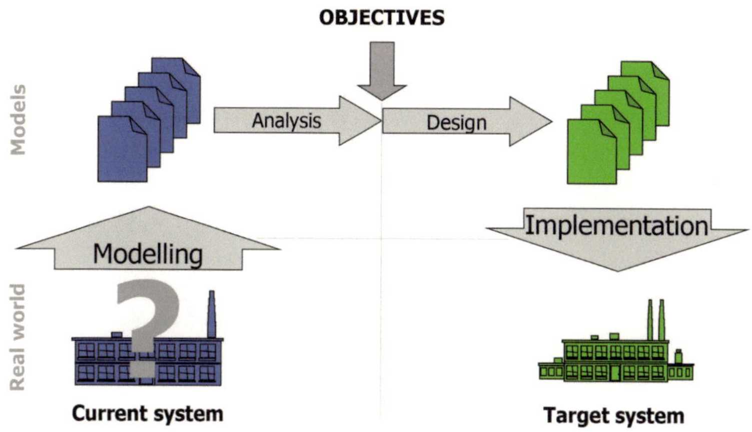

Figure 8.
The general principle of conceptualisation of enterprise modelling.

existing state to the model of the future one and, 3. *implementation*: passage from the model of the future state to the new reality). It is easy to see the analogy with what is proposed in the migration plan.

This principle also explains why the sequence followed by this level is opposite to that of the strategic orientation level. In this one, the first step concerned the formalisation of the target, with the existing state being dealt with afterwards. This sequence makes it possible to link all this level to the strategic analysis of the company. Within the framework of the migration plan level, the existing state is processed (modelled) first. This enables the model of the first step to be developed on its representation in terms of performance from the previous level (**Figure 7**) but also from the analysis carried out on the basis of the models of the existing state (first action: current state modelling).

4.3 The projects portfolio level

As for the previous level, the two time milestones structuring this level are the current state and the first step. The difference with the previous level is that here the two milestones are expressed in the form of projects. The change of modes of expression reflects the desire to move from a static vision (the models represent the states of the system) to a dynamic vision (the actions that need to be taken to move from one state to another). That is why the projects portfolio is called "To-do" **Figure 9**, in comparison with the "To-be" of the upper level.

Moving from a model to a list of projects is not an obvious task. This is why the last activity of the migration plan was to propose an action plan. Then, this action plan is the link between the models and the projects. However, the action plan was mainly aimed at analysing what the envisaged migration entails. That is why it was not very precise in terms of timing or resources mobilised. The project portfolio level must fill this gap in the sense that all the elements that make up a real project must now be defined.

The three main activities that must be carried out within this level are as follows.

1. *Current state evaluation*. The objective of this activity is to analyse the progress and results of recent projects i.e., those belonging to the previous version of the projects portfolio. This analysis has a double purpose. Firstly, it is to verify that the projects that have just been carried out have achieved their objectives. If this is not the case, corrective or compensatory actions in the form of projects

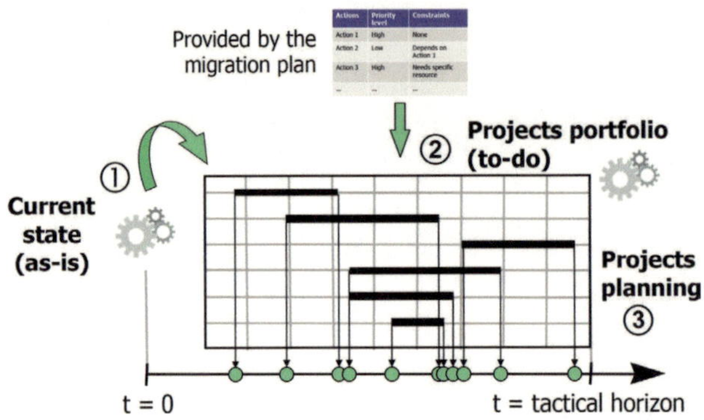

Figure 9.
The activities of the projects portfolio level.

will have to be integrated into the new projects portfolio. This assessment reflects the fact that two successive instances of the projects portfolio are not independent. It also corresponds to some extent to the Check and Act phases of the PDCA. The second reason for this evaluation is the fact that some projects may not have been carried out within the tactical horizon of the previous project portfolio, contrary to what should have been the case. This may be due either to a decision to run a project beyond this horizon, or to the fact that a project has been postponed for various reasons. In the end, this activity makes it possible to know perfectly the state of progress decided the previous time and to take this state into account for the definition of the new projects portfolio.

2. *Projects portfolio definition*. This activity is central at this level as it is the one that defines the projects portfolio. This is built on the basis of the action plan provided by the higher level. It is clear that the transition from actions to projects is not based on a bijective relationship: several actions can be grouped together to form a single project and, conversely, one action can lead to several projects. The latter case is classic and corresponds to a secondary need arising from the initial project. For example, a change in a management function (initial project) leads to the need to launch a computerisation project and a project to train the managers concerned (secondary projects). The difference between the actions plan and the projects portfolio is that this level takes into account various constraints that had not been considered at the higher level: financial resources, availability of human resources, negotiation with solution providers, etc. The second element to be taken into account is the evaluation carried out by the previous activity: definition of corrective or compensatory activities and integration into the portfolio of ongoing projects. The importance of taking this assessment into account is clear: ongoing projects consume resources that will therefore be unavailable for new projects and they may constitute precedence constraints for new projects.

3. *Projects planning*. There are therefore as many activities as there are defined projects. The tasks to be defined and planned are standard:

    - Drawing up specifications: definition of technical specifications in relation to the models provided by the Migration plan.

    - Design or acquisition: development or purchase on the market of the solutions identified during the previous phase.

- Implementation and integration of the components developed or purchased.

The main elements to be taken into account are also standard: positioning of projects over time, conditions of precedence between projects, organisation of the company's internal resources, and triggering the involvement of external resources.

The horizon of this management is variable since it corresponds to the duration of the project concerned. It falls between two time milestones corresponding to the beginning of the project and its end. All these milestones constitute a sequence of events that set the pace of the projects portfolio's evolution (**Figure 9**).

It is important to find the best compromise between independence in the management of each project and overall coordination within the projects portfolio.

Figure 9 shows these activities.

Finally, this level deals with project management with classical constraints and concepts. The important point is the existence of several concurrent and coordinated projects.

5. Argumentation

The proposed approach highlights several aspects that contribute to the competitiveness of enterprises. The main ones are listed here. On the other hand, taking the approach to its ultimate conclusion presupposes that the company develops self-assessment capacities. We will come back to this point in the second part.

5.1 Competitive aspects

Performance evaluation. The approach emphasises the notion of performance. It is a major element to be integrated into the management of modern companies because, in order to manage their evolution, companies need to evaluate their performance level (actual state) and compare it with a projected state defined in relation to the economic environment. This expected target with performance evaluation and the path to achieve is also evoked in A3 approach of Lean when targets are evoked to allow easier projection of corrective actions. Faced with competitive pressure, many companies have moved in this direction in recent decades. Nevertheless, knowing how to measure performance and how to choose the corresponding indicators is not yet a talent that all companies still possess. This is why many methods have been proposed to help companies move in this direction [31, 32].

Industrial strategy. Talking about performance also means talking about strategy, because it is strategy that allows to clearly define the performance to be monitored. Moreover, an improvement project requires a clear definition of the target to be reached through the formulation of an industrial strategy. This first requires the development of a strategic vision/target to ensure coherence and synergy between all the improvement projects carried out. This argument is not shared by all companies. Obviously, large groups build strategic plans but many SMEs do not for many reasons [33]. Whatever the arguments, the proposed approach encourages the definition of a strategy before any intention of evolution.

Models. To propose modelling is to encourage companies to acquire the means to know themselves. Models do not bring new knowledge about the company, but they allow it to be expressed, standardised and exchanged. As mentioned in [34], to model is to externalise knowledge. Self-organising means choosing one's trajectory and adapting accordingly; it presupposes being able to generate symbolic information, i.e. information about oneself [35, 36]. Models contribute to this. Also, pushing companies to model themselves means pushing them to know perfectly and permanently how they run and the behaviour of each of their components and to identify which part of the structure needs to be improved or changed. It means enabling them to be autonomous in managing their evolution.

Motivation. Employee motivation is linked to the significance of the work [37]. In terms of change management, this is expressed by the knowledge of the target (where the company is going) and the possibility of frequently see the results of projects. Then, proposing an approach organised in small projects that allow to reach a step of evolution, itself positioned in relation to a long-term target, allows everyone in the company to appreciate the path proposed and the results obtained. It is also important that employees be involved in the approach as much as possible, which is what continuous improvement and most enterprise modelling methods propose. Ali et al. [38] mentions the lack of training and planning as barriers to Lean projects implementation. These aspects have to be systematically taken into account in the projects portfolio.

5.2 Self-evaluation and learning capabilities

The current state (as-is) must be expressed at each level, in the three proposed forms: performances, models and ongoing projects. As the approach is presented, this expression is based on a fully-fledged activity at the three levels of evolution management, i.e. this state is reconstructed each time. This reconstruction can be carried out by the company itself or by relying on the services of an external company, which is often the case.

Pushing the logic to its ultimate conclusion means thinking in terms of *internalisation* and *continuity*.

Internalisation reflects the fact that the company must be able to do this on its own. Indeed, knowing how to evaluate its performance, model its own operations and monitor its projects are not these skills that every well-organised company should have within it? Just as it is normal for the company to turn to external service providers for design activities (because it may not have the necessary skills in IT, workstation organisation, etc.), it is also necessary for it to be able to express its current state.

Continuity is the principle that the company should not have to reconstruct its current state at each step of the process but should be able to know it at every moment. As regards the strategic orientation level, this means implementing a system of performance indicators (performance monitor) that is updated as often as possible and that can be adapted if strategic orientations require a change of indicators. For plan level migration, this means that the company has its own models and that there is someone responsible for updating them each time a change is noticed. By analogy with the technical data that the company necessarily possesses for its technical activities, this set is called organisational data here. Finally, for projects portfolio level, it means following and monitoring the evolution projects (ongoing projects portfolio), which in general is integrated into the company's operations and does not pose any problems. These three elements are grouped together in a set entitled "Enterprise monitoring and documentation". Finally, continuity reflects the obvious fact that in order to evolve continuously, the enterprise must be able to evaluate itself continuously.

In conclusion, the approach proposed here leads to advocate a vision of the enterprise that takes its evolution in hand and that provides itself with the means to constantly learn about and evaluate itself. In this way, the evolution management participates to the development of learning organisations [39, 40].

Figure 10 summarises this vision and shows the main activities.

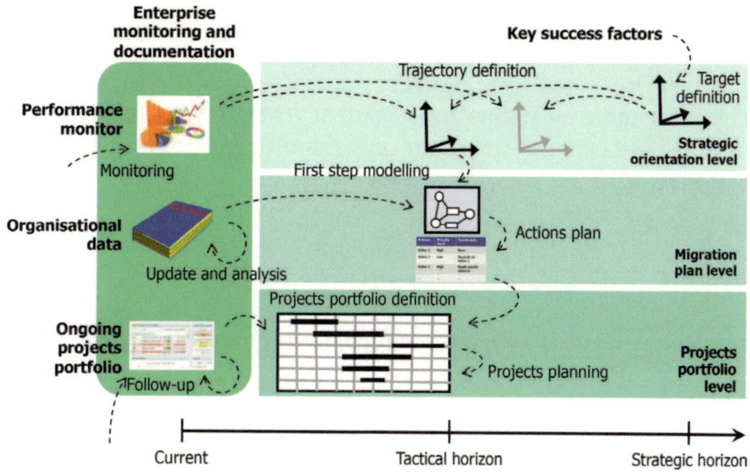

Figure 10.
The organisation of evolution in the self-evaluated, learning enterprise.

5.3 Evolution management and continuous improvement

The evolution management system entails many aspects consolidating the PDCA approach, well known and used in large groups and even SMEs to sustain quality. Even though strategy, as quoted in section 5.1 is not obviously formalised by SMEs because of their dependencies to big groups, they often use and admit efficiency of PDCA vision or DMAIC (often tightly linked to project approach and can also be assimilated to PDCA cycle). Indeed, PDCA is the fundament of continuous improvement because of the value given to the "Act" step to ensure continuous action on systems to make a progress. In the vision presented here, the actions to carry out are in step "Act" of the PDCA but are no more only corrective actions after "Check" step. They represent also new proactive ideas and prospective plans to improve the whole existing projects system regarding the "output" and "knowledge" got from ongoing projects portfolio and migration plan.

The evolution management reminds the importance for the company to continually formalise and display the targeted performances. The performance objectives are tightly linked to the defined "Strategy" that can be revealed in "Plan" step of PDCA. Updating with "performance targets" planned by company strategy is the potential inducer of "could be" situations.

Concerning migration plan, PDCA and Lean highlight that, whatever modelling approach considered, the "added value" is always the main concept to undertake to keep "efficient" model with the required added values processes, the expected relevant data, the prior decisions and the accurate organisations.

To model the current state (As-is), we should remind that the use of various instruments available to an analyst to build the model as consultation of documents, analysis of computer application screens, field observations and interviews are such many elements absolutely necessary to deal with "reliable" data. From Lean point of view, any process has to be produced respecting "Jidoka" notion which means ensuring the quality "at the source". The current state modelling is critical step that should be made as reliable as possible to avoid wasting times and retro-corrective actions. The more the system is reliably represented the better the "could be" system can be achieved in good conditions. So Jidoka, principle coming from Lean management, is an efficient support for the organisational data sustainability.

Lean practices and Continuous improvement are indubitably the result of human forces, company strategy and collective efforts. By the way, the motivation and involvement of the team project evoked previously is an important part defended by Lean and continuous improvement. Then, evolution management, if well described and explained to the team, is significantly able to strengthen the "Do" step of PDCA.

6. Conclusions

The approach presented here aims at repositioning the enterprise modelling approach in the context of continuous evolution, better able to respond to a turbulent economic environment.

Within this framework, it emerges that many tools and approaches are involved in the reengineering and improvement of companies: strategy, performance measurement, modelling, projects and the whole toolbox of Lean Management and continuous improvement. The approach presented here is an opportunity to bring these approaches closer together: strategy and performance measurement at the top level, Lean models and tools at the central level and projects at the operational level.

The ultimate goal is to take advantage of the benefits of all these approaches. For example, Lean insists on short projects, anchored in practice and part of a continuous improvement; enterprise modelling allows to document the company, to share knowledge and to propose a systemic vision.

Finally, the approach proposed here opens important perspectives concerning the integration of enterprise reengineering approaches.

Author details

Bruno Vallespir* and Anne Zouggar-Amrani
Univ. Bordeaux, CNRS, IMS, UMR 5218, Talence, France

*Address all correspondence to: bruno.vallespir@ims-bordeaux.fr

References

[1] Doumeingts G, Vallespir B. Marcotte F. A proposal for an integrated model of manufacturing system: application to the re-engineering of an assembly shop. Control Engineering Practice. 1995;3(1):59-67. DOI: 10.1016/0967-0661(94)00065-O

[2] Vernadat F. Techniques de Modélisation en Entreprise : Application aux Processus Opérationnels. Paris: Economica; 1999. 129 p.

[3] Vernadat F. Enterprise modelling: Research review and outlook. Computers in Industry, 2020;122:103265. DOI: 10.1016/j.compind.2020.103265

[4] Vallespir B, Braesch C, Chapurlat V, Crestani D. L'intégration en modélisation d'entreprise : les chemins d'UEML. In Proceedings of the 4th Conférence Francophone de Modélisation et Simulation (MOSIM); 23-25 April 2003; Toulouse, France

[5] Vallespir B, Ducq Y. Enterprise Modelling: from early languages to models transformation. International Journal of Production Research. 2018;56(8):2878-2896. DOI: 10.1080/00207543.2017.1418985

[6] AMICE. CIMOSA: Open Architecture for CIM. Berlin: Springer; 1993. 234 p.

[7] Kosanke, K, Vernadat F, Zelm M. CIMOSA: enterprise engineering and integration. Computers in Industry, 1999;40(2-3):83-97. DOI: 10.1016/S0166-3615(99)00016-0

[8] Vallespir B, Merle C, Doumeingts G. GIM: a technico-economic methodology to design manufacturing systems. Control Engineering Practice. 1993;1(6): 1031-1038. DOI: 10.1016/0967-0661(93)90014-I

[9] Williams TJ, Ernus P, Brosvic J, Chen D, Doumeingts G, Nemes L, Nevins JL, Vallespir B, Vlietstra J, Zoetekouw D. – Architectures for integrating manufacturing activities and enterprises. Computers in Industry. 1994;24(2-3):111-139. DOI: 10.1016/0166-3615(94)90016-7

[10] GERAM. GERAM: Generalised Enterprise Reference Architecture and Methodology. Version 1.6.1, IFIP–IFAC Task Force on Architectures for Enterprise Integration; 1999

[11] Panetto H, Mayer F, Lhoste P. Unified Modelling Language for meta-modelling: towards Constructs definition. In Proceedings of the 10th symposium Information Control in Manufacturing (INCOM); 20-22 September 2001; Vienna, Austria

[12] Chen D, Vallespir B, Doumeingts G. Developing an unified enterprise modelling language (UEML) – Roadmap and requirements. In Proceedings of the 3rd IFIP Working conference on infrastructures for virtual enterprise (PROVE); 1-3 May 2002; Sesimbra, Portugal

[13] Roque M, Vallespir B, Doumeingts G. Interoperability in enterprise modelling: Translation, elementary constructs, meta modelling and UEML development. Computers in industry. 2008;59(7):672-681. DOI: 10.1016/j.compind.2007.12.017

[14] Bézivin J. From Object Composition to Model Transformation with the MDA. in Proceedings of the International Conference on Technology of Object-Oriented Languages (TOOLS); 29 July-3 August 2001; Santa Barbara, California; 350-354. DOI: 10.1109/TOOLS.2001.10021

[15] Blanc X, Salvatori O. MDA en action: Ingénierie logicielle guidée par les modèles. Paris: Eyrolles; 2011. 298 p.

[16] Estefan JA. Survey of model-based systems engineering (MBSE)

methodologies. Incose MBSE Focus Group 25; 2007

[17] Gabriel E, The lean approach to project management. International Journal of Project Management. 1997;15(4):205-209. DOI: 10.1016/S0263-7863(96)00066-X

[18] Sousa P, Tereso A, Alves A, Gomes L. Implementation of project management and lean production practices in a SME Portuguese innovation company. Procedia Computer Science. 2018;138:867-874. DOI: 10.1016/j.procs.2018.10.113

[19] Tenera A, Pinto LC. A Lean Six Sigma (LSS) project management improvement model. Procedia - Social and Behavioral Sciences. 2014;119:912-920. DOI: 10.1016/j.sbspro.2014.03.102

[20] Cheng CY, Chang PY. Implementation of the Lean Six Sigma framework in non-profit organizations: A case study. Total Quality Management & Business Excellence. 2012;23(3-4):431-447. DOI: 10.1080/14783363.2012.663880

[21] Veres C. Conceptual Model for Introducing Lean Management Instruments. Procedia Manufacturing. 2020;46:233-237. DOI: 10.1016/j.promfg.2020.03.034

[22] Ballé M, Jones D, Chaize J, Fiume O. La stratégie Lean : Créer un avantage compétitif, libérer l'innovation, assurer une croissance durable en développant les personnes. Paris: Eyrolles; 2018. 352 p.

[23] Hohmann C. Lean Management: Outils - Méthodes – retours d'expériences - Questions/réponses. Paris: Eyrolles; 2012. 424 p.

[24] Masuti PM, Dabade UA. Lean manufacturing implementation using value stream mapping at excavator manufacturing company. Materials Today: Proceedings. 2019;19(2):606-610. DOI: 10.1016/j.matpr.2019.07.740

[25] Amrani A, Ducq Y. Lean practices implementation in aerospace based on sector characteristics: methodology and case study. Production Planning & Control. 2020;31(16):1313-1335. DOI: 10.1080/09537287.2019.1706197

[26] Song MH, Fischer M. Daily plan-do-check-act (PDCA) cycles with level of development (LOD) 400 objects for foremen. Advanced Engineering Informatics. 2020;44:101091. DOI: 10.1016/j.aei.2020.101091

[27] Doumeingts G, Kleinhans S, Malhéné N. GEM TIME: a proposal for an evolution management methodology, in Proceedings of Advanced Production Management Systems (APMS); 4-6 November 1996; Kyoto, Japan.

[28] Malhéné N. Gestion du processus d'évolution des systèmes industriels – conduite et méthode [thesis]. University Bordeaux 1; 2000.

[29] Doumeingts G. GEM: GRAI evolution method: a case study. International Journal of Technology Management. 2001;22(1-3):189-211. DOI: 10.1504/IJTM.2001.002961

[30] Zanettin M. Contribution à une démarche de conception des systèmes de production [thesis]. University Bordeaux 1; 1994.

[31] Ravelomanantsoa M, Ducq Y, Vallespir B. A state-of-the-art and comparison of approaches for performance measurement systems definition and design. International Journal of Production Research. 2019;57(15-16):5026-5046. DOI:10.1080/00207543.2018.1506178

[32] Ravelomanantsoa M, Ducq Y, Vallespir B. General enterprise performance measurement architecture. International Journal of Production

Research. 2020;58(22):7023-7043. DOI:10.1080/00207543.2019.1692158

[33] Wang C, Walker EA, Redmond JL. Explaining the lack of strategic planning in SMEs: The importance of owner motivation. International Journal of Organisational Behaviour. 2007; 12(1):1-16.

[34] Lillehagen F, Krogstie J. Active Knowledge Models and Enterprise Knowledge Management. In Proceedings of the International Conference on Enterprise Integration and Modeling Technology (ICEIMT); 24-26 April 2002; Valencia, Spain: Kosanke K, Jochel R, Nell JG, Ortiz Bas A editors. Enterprise Inter- and intra-organizational integration, Berlin: Springer 2002, p. 91-99. DOI:10.1007/978-0-387-35621-1_43

[35] Boulding, KE. General Systems Theory-The Skeleton of Science. Management Science. 1956;2(3):197-208.

[36] Le Moigne JL. La théorie du système général. Théorie de la modélisation. Paris: PUF; 1977. 339 p.

[37] Dwivedula R, Bredillet C. Profiling work motivation of project workers. International Journal of Project Management. 2010; 28(2):158-165. DOI:10.1016/j.ijproman.2009.09.001

[38] Ali SM, Hossen MA, Mahtab Z, Kabir G, Paul SK, Adnan ZUH. Barriers to lean six sigma implementation in the supply chain: An ISM model. Computers and Industrial Engineering. 2020;149:106843. DOI: 10.1016/j.cie.2020.106843

[39] Argyris C, Schön DA. Organizational learning: A Theory of Action Perspective. Reading: Addison Wesley; 1978. 344 p.

[40] Hayes RH, Wheelwright SC, Clark KB. Dynamic manufacturing – Creating the learning organization. New York: The free press; 1988. 400 p.

Chapter 8

Single Minute Exchange of Dies: Classical Tool of Lean Manufacturing

Yash Dave

Abstract

Effective utilization of the resources is the need of an hour particularly when it comes to the manufacturing industry. It is having a paramount importance to have a proper utilization of the resources, on the same line in any manufacturing industries to reduce the setup time is also one of the ways to do so. Single Minute Exchange of Dies (SMED) is one of the classical method which is normally used to reduce the setup time. In this technique complete videography of the existing changeover is done and then by analyzing it waste activities identified and other improvement plant has been done in each iteration. The chapter also showcases the SMED technique applications in a gear industry. Remarkable resources and results have been achieved through the implementation of classical tool of Lean manufacturing is made.

Keywords: SMED, rapid change over, waste reduction, lean tool

1. Introduction to SMED

Modern Times every manufacturer wants to produce product in various varieties. This can only be accomplished when we have a shorter setup times in order to change the product from one variety to another. SMED means no setup time should be more than nine minutes. SMED is a systematic procedure to reduce setup time by eliminating waste.

2. History of SMED

It is a special technique of a Lean manufacturing system. It is also an integral part of Toyota production system. It is basically devised by the Shingo. Normally the setup time in the industries at very high and this creates a major obstacle in order to do the rapid change over. This can be easily and effectively compensated by SMED initially owner at Toyota able to reduce the setup time from 24 hours to few minutes. SMED is also considered as one of the tool to increase the flexibility. Shingo shows that this method can be applied to any type of industries. In order to provide the elements it details documentation procedure Shingo has written a classical book "A Revolution in Manufacturing: The SMED System". It is one of the first and most valuable resources available on the SMED. In India also, the application of SMED in all types of industries have been reported in academic literature [1]. Discussed the SMED methodology in garment manufacturing industry and reported considerable reduction in delay arising out of machine setting time, batch setting time and demonstration delay [2]. Presented the SMED application in automobile industry and reported that changeover time was reduced to 24.5% and considerable improvement in the productivity [3]. Showcased the experimental work carried out at automobile industry, located in Maharashtra, India. Implementation of SMED tool resulted in reduction of setup time by 82.44% and tool change time by 44.21%. This helped to produce additional 23 jobs/shift with same input [4]. Implemented the SMED tool and reported the significant savings with minimum investment and also highlighted the importance of safety and ergonomic factors during die exchange [5]. Presented a case study of the application of SMED in medicinal product based industry and reported remarkable achievement in the form of ability to reduce the batch size which can compensate the market demand fluctuations and also able to reduce the storage of medicinal products that are very closely related to expiration time (expired time).

3. SMED terms and its elements

Following are the most commonly terms associated with the SMED nomenclature. Basic elements of the entire SMED system have also been shown in **Figures 1**, 2 and 3:

- Changeover: The process of switching from the production of one particular product or model to another on a machine.

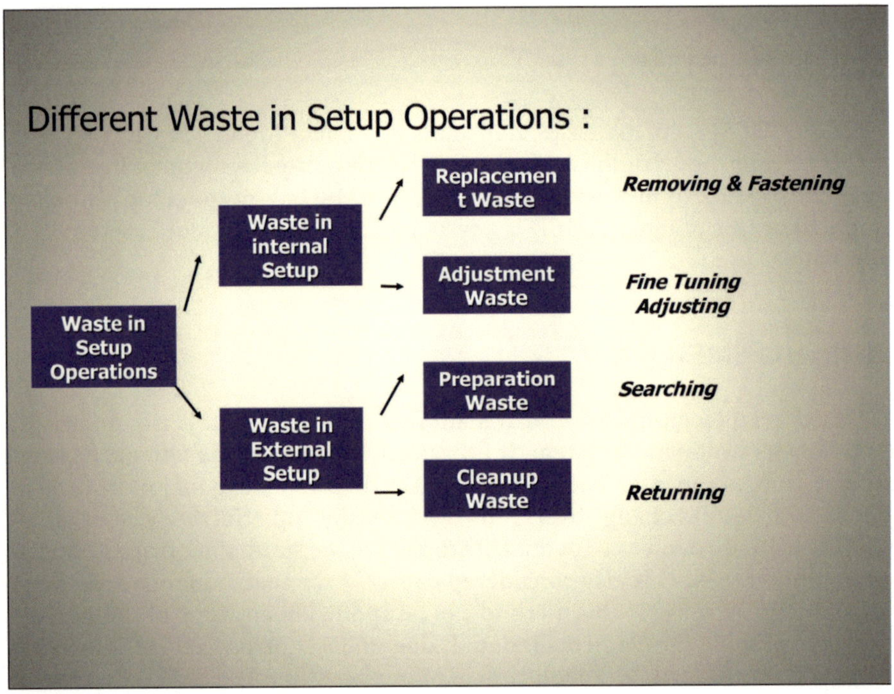

Figure 1.
Different waste in setup Operations.

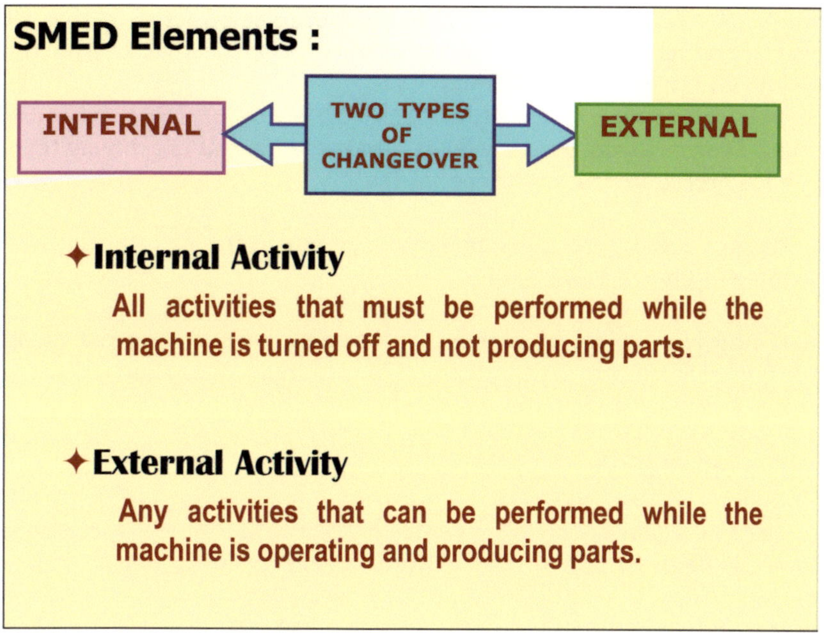

Figure 2.
Basic SMED elements (1).

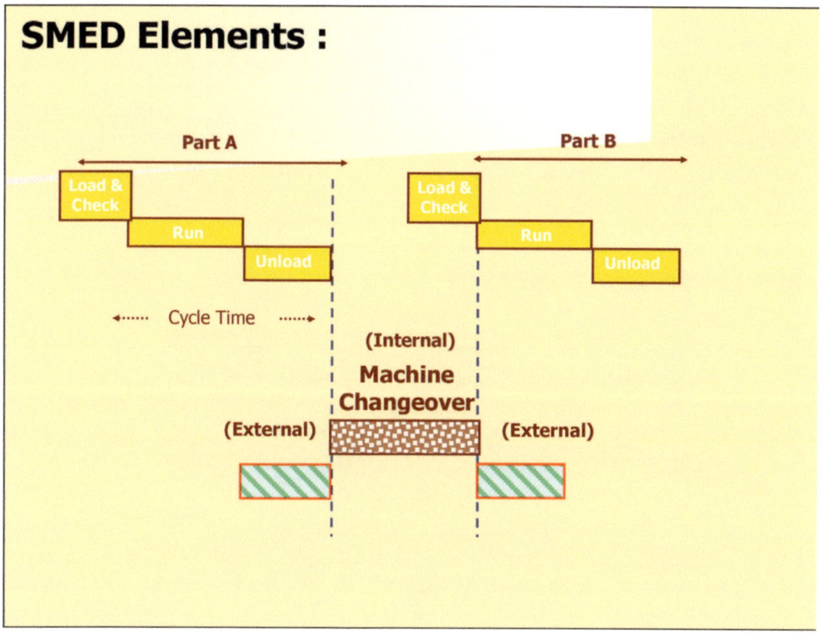

Figure 3.
Basic SMED elements (2).

- Downtime: Production time lost due to unwanted stoppages.

- External Setup activities: Those activities or tasks which can be done while the machine is still running are known as external setup activities.

- Internal Setup activities: Those activities or tasks which can be done while the machine is shut down are known as internal setup activities.

- Non-Value Added Activities: The resource consumed on activities that add costs but no value to an item from the customer's perspective. These activities are known as non-value added activities.

- Shigeo Shingo: Toyota person who has dedicated himself in evolving the detailed procedure and aspects of SMED.

- Waste: Any activity that consumes resources but creates no value for the customer is known as waste.

- 5S: A basic lean tool which is primarily the collection of five Japanese language words used for improving productivity in any organization and industry

4. SMED methodology

The entire methodology and step by step procedure of the SMED technique has been shown in **Figure 4**. It is primarily consists of the following steps:

4.1 Observe the current methodology (changeover)

In this step the entire existing changeover procedure is recorded through video camera

4.2 Separate the internal and external activities

In this step the complete changeover is analyzed frame by frame by the entire SMED team members and all the activities is separated in to three categories namely internal activities, external activities and waste activities.

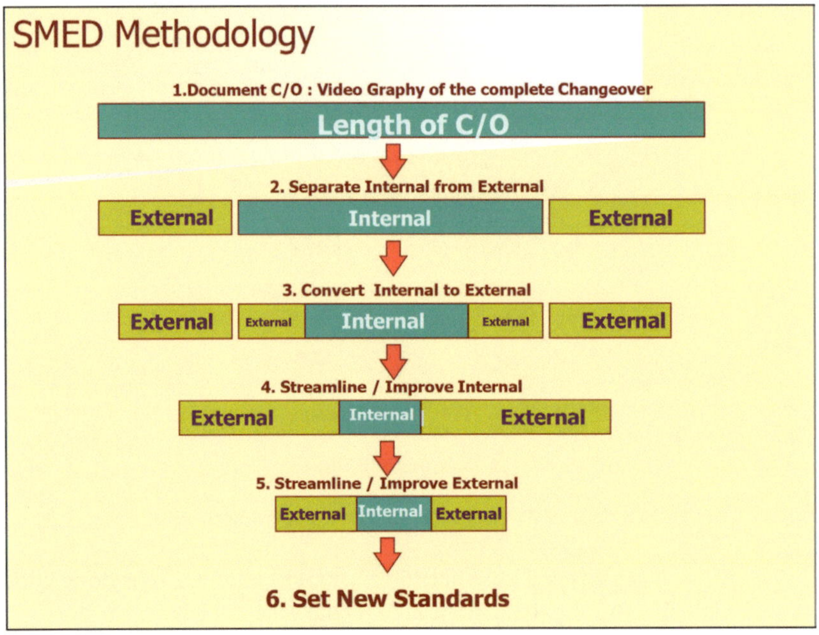

Figure 4.
SMED methodology.

4.3 Convert (where possible) internal activities into external ones

After segregation of the activities, SMED team converts as possible as internal activities in to external activities by implementing the rigid improvement plan and kaizen activities

4.4 Training to all operators

If any improvement plan requires the change in working of the operator's method then it is most important to impart the rigorous training of the change procedure to the entire operator team not only to teach them about the new procedure but also to inculcate the trust among them about the new changed procedure.

4.5 Do it all again

As the concept of SMED tells us to achieve the setup time in single digit minute time but in practical conditions it is not possible to achieve this goal in single iteration. It needs lots of iteration in order to achieve the goal. So it is always required that keep doing the entire procedure until and unless goal is achieved.

4.6 Set new standards

After everyiteration, new standard should be setup by the team and it should also be treated as benchmark in order to haveworld class manufacturing level

5. Example of SMED technique application

The implementation of the SMED technique is little bit complicated but it can be implemented effectively if structured methodology is adopted. In this example SMED methodology is applied at the gear manufacturing unit for reducing setup time of shaving machine of a Gear company cell producing a large diameter gears (**Figure 5**) and following steps are followed:

5.1 Understanding the current situation

The current situation and layout of the gear manufacturing cell is shown in **Figure 5**. Here in this step the complete videography of the changeover procedure of a set up time of shaving machine is done and by watching and analyzing the videography of the changeover procedure the changeover sheet has been prepared as shown in **Table 1**.

5.2 Identification of waste or abnormalities

As shown in the **Table 2** there is lot of waste activities in the complete setup procedure of the shaving machine. Total time consumed by the waste activities is 44 minutes approximately. In first iteration itself, all the waste activities are eliminated.

5.3 Analysis of the causes

On analyzing the complete changeover sheet and various brainstorming session with all stakeholders are organized. Following prominent causes which were surfaced during this exercise was as follows (**Figure 6**).

i. Training of set up change procedure was not given properly to the setup personnel and there were no Master personnel of setup for particularly Shaving Machine.

ii. 5S of setup trolley is not properly organized.

iii. Lot of time is being wasted due to lead andprofile (L & P) is not ok at the first time itself.

To analyze the cause of Lead and Profile problem more deeply, a detailed study has beencarried out to find the root cause of the problem which is shown in the **Figure 7.**

5.4 Improvement plan and implementation

After finding the root cause of all waste activities found in first iteration a time bound improvement plan has been made and all the improvement has been made as per the plan which is shown in **Figure 8**.

5.5 Result and Kaizen done in first iteration

In this way all the wasteactivities which has found in first iteration completely eliminated. The major result achieved by this exercise is that by eliminating all waste activities which are acting as a hindrance to convert the internal activities in to the external activities can now be taken up in second iteration. By doing the first iteration of the reducing set up time cycle the total 44 minutes have been saved.

Kaizen done: In this complete exercise following kaizen has been done which is listed as follows: **Kaizen 1**: As shown in **Figure 7**, cutter head angle is the responsible factor for wrong lead profile because during setup (first iteration) it was wrongly set as per the requirement to suit the shaving cutter. To solve this issue Cutter head angle is made mandatory check point in the standard operating procedure.

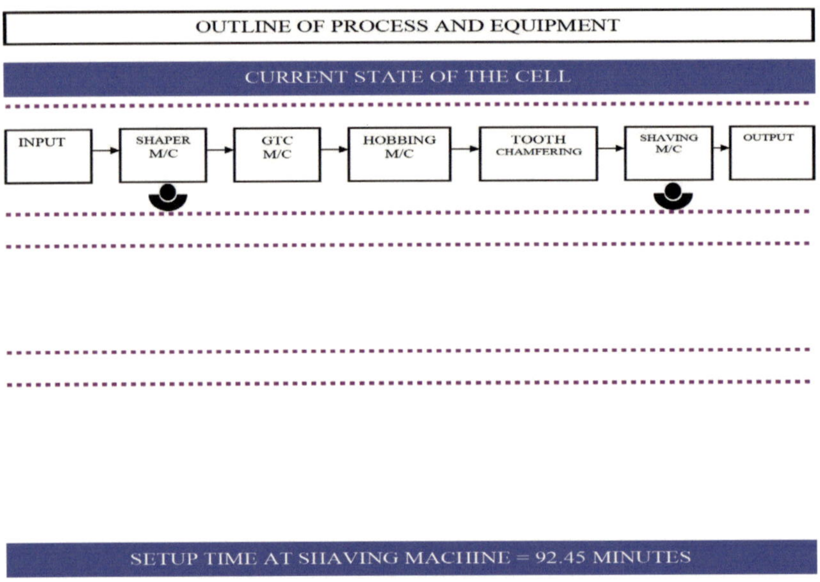

Figure 5.
Outline of process and equipment.

CHANGEOVER OPERATIONS ANALYSIS CHART							
TOTAL MINUTES 92.45		DEPT. PRODUCTION		DESCRIPTION OF CHANGEOVER From MODEL 808 to 809			
		MACHINE(S) SHAVING		SET-UP PERSONNEL Vishesh			
NO.	CHANGEOVER ELEMENTS	RUNNING TIME	ELEMENT TIME (sec.)	CHANGEOVER CATEGORIES INTERNAL	EXTERNAL	WASTE	IMPROVEMENT PLAN
1	Loosening cutter arbor guide bolts.	1:13:40	1:13:40				
2	Taking out guide	2:00:00	0:46:20				
3	Cleaning cutter housing face	2:15:00	0:15:00				
4	Loosening locknut locking bolts	2:40:00	0:25:00				
5	Loosening locknut & taking out cutter	5:54:00	3:14:00				
6	Putting 808 cutter in box	7:00:00	1:06:00				
7	Cleaning cutter area & arbor face	8:02:00	1:02:00				
8	Cleaning 809 cutter	8:30:00	0:28:00				
9	Mounting cutter on arbor (Spacer/cutter/spacer/locknut) cleaning every part before assembly (Activity stopped due to grub screw damage)	11:32;00	3:02:00				
10	Changing helix angle (Unlocking cutter head/ change angle/ lock cutter head)	14:47:00	3:15:00				
11	Locking cutter (after procuring grub screw)	17:09:00	2:22:00				
12	Mounting arbor guide (after cleaning face & ID)	22:52:00	5:43:00				
13	Reading QAP for DOB size	23:28:00	0:36:00				
14	Changing crowning angle	26:03:00	2:35:00				
15	Setting depth for 0809	28:03:00	2:00:00				

CHANGEOVER OPERATIONS ANALYSIS CHART							
TOTAL MINUTES 92.45		DEPT. PRODUCTION		DESCRIPTION OF CHANGEOVER From MODEL 808 to 809			
		MACHINE(S) SHAVING		SET-UP PERSONNEL Vishesh			
NO.	CHANGEOVER ELEMENTS	RUNNING TIME	ELEMENT TIME (sec.)	CHANGEOVER CATEGORIES INTERNAL	EXTERNAL	WASTE	IMPROVEMENT PLAN
16	Checking component centre	28:35:00	0:32:00				
17	Setting depth for 0809	29:51:00	1:16:00				
18	Setting length of cut	31:55:00	2:04:00				
19	Increasing carrier height	34:02:	2:07:00				
20	First pc loading in mandrel	34:51:00	0:49:00				
21	Cycle ON	37:16:00	2:25:00				
22	Comp stuck while unloading	40:10:00	2:54:00				
23	Size checking (not OK)	41:10:00	1:00:00				
24	Reloading and setting the size	43:22:00	2:12:00				
25	Checking size (OK)	43:42:00	0:20:00				
26	Given to std. room for L&P checking (LP and OK)	55:39:00	11:57:00				
27	Lifting table for L&P correction	58:00:00	2:21:00				
28	Cycle ON Idle	60:20:00	2:20:00				
29	Cycle ON with component	61:21:00	1:01:00				
30	Given to std. room for L&P checking (L&P not OK)	80:33:00	19:12:00				
31	Setting helix angle lead correction	84:40:00	4:07:00				
32	Cycle ON Idle	85:36:00	0:56:00				

CHANGEOVER OPERATIONS ANALYSIS CHART							
TOTAL MINUTES 92.45		DEPT. PRODUCTION		DESCRIPTION OF CHANGEOVER From MODEL 808 to 809			
		MACHINE(S) SHAVING		SET-UP PERSONNEL Vishesh			
NO.	CHANGEOVER	RUNNING	ELEMENT	CHANGEOVER CATEGORIES			IMPROVEMENT PLAN
	ELEMENTS	TIME	TIME (sec.)	INTERNAL	EXTERNAL	WASTE	
33	Cycle ON with component	86:27:00	0:51:00				
34	Given to std. room for L&P checking (L&P OK)	92:45:00	6:18:00				

Table 1.
Changeover sheet.

Activity No.	Waste Activities	Time in Sec (2650 Sec)
3	Cleaning cutter housing face	15
16	Checking component centre	32
17	Setting depth for 0809	76
19	Increasing carrier height	127
22	Component stuck while unloading	174
27	Lifting table for L&P correction	141
28	Cycle ON Idle	140
29	Cycle ON with component	61
30	Given to std. room for L&P checking (L&P not OK)	1152
31	Setting helix angle lead correction	247
32	Cycle ON Idle	56
33	Cycle ON with component	51
34	Given to std. room for L&P checking (L&P OK)	378

Table 2.
Identified waste activities.

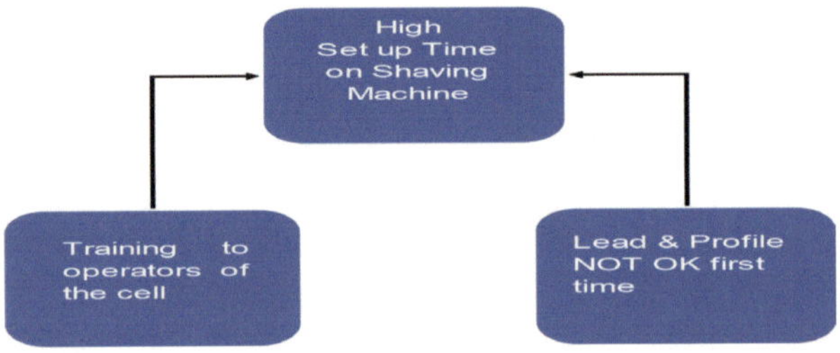

Figure 6.
Major causes of high setup time.

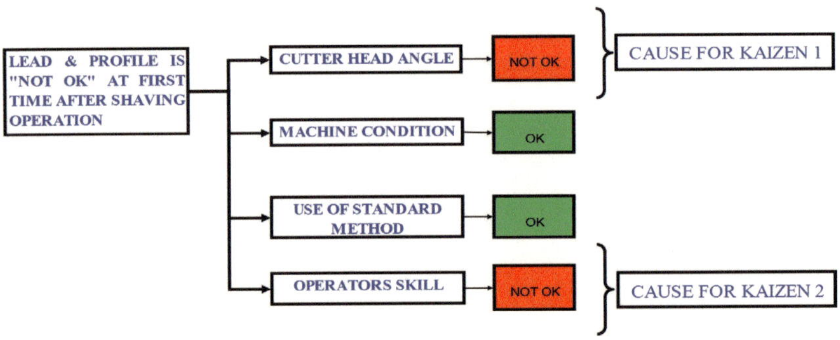

Figure 7.
Analysis of the major causes.

Kaizen 2: As shown in **Figure 7**, the poor operator skill of setup change procedure is responsible factor for the cause of this Kaizen. To solve this issue following improvement activities has been done:

a. One operator is made in charge of setup changeover and he has been given rigorous training for standardized operating procedure for the setup changeover and provided with written documents.

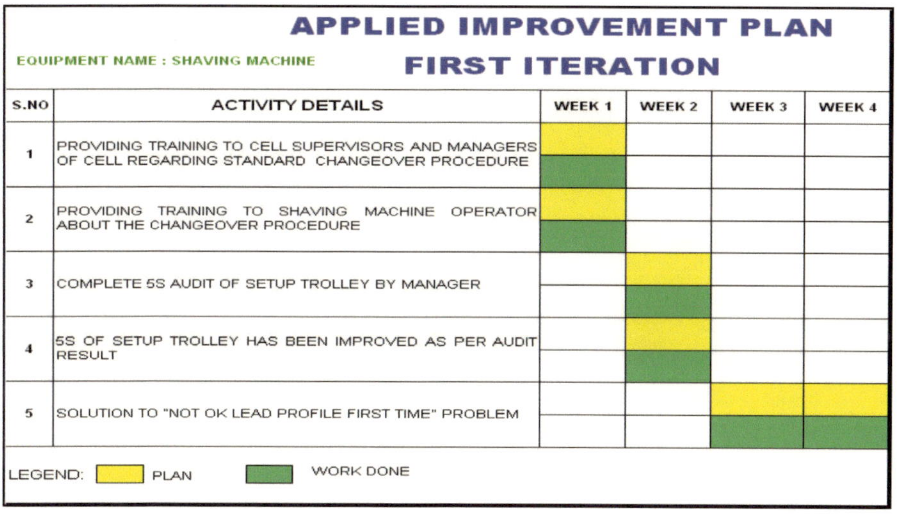
APPLIED IMPROVEMENT PLAN
FIRST ITERATION

EQUIPMENT NAME : SHAVING MACHINE

S.NO	ACTIVITY DETAILS	WEEK 1	WEEK 2	WEEK 3	WEEK 4
1	PROVIDING TRAINING TO CELL SUPERVISORS AND MANAGERS OF CELL REGARDING STANDARD CHANGEOVER PROCEDURE				
2	PROVIDING TRAINING TO SHAVING MACHINE OPERATOR ABOUT THE CHANGEOVER PROCEDURE				
3	COMPLETE 5S AUDIT OF SETUP TROLLEY BY MANAGER				
4	5S OF SETUP TROLLEY HAS BEEN IMPROVED AS PER AUDIT RESULT				
5	SOLUTION TO "NOT OK LEAD PROFILE FIRST TIME" PROBLEM				

LEGEND: PLAN WORK DONE

Figure 8.
Improvement plan and its implementation.

b. In charge of setup changeover has been provided with a complete Setup trolley with rearranged 5S.

c. A skill matrix is being developed for each operator so that each operator of the cell would be capable to perform setup changeover as per the standard operating procedure.

6. Conclusion

In true sense, SMED is a classical tool of lean manufacturing. As described above there are some apparent benefits for any industry by implementing a SMED tool.

1. It automatically creates lot of extra capacity in any sale our industry this provides a platform for any industry to produce more items that means effective utilization of resources.

2. Since during implementation of SMED always a restricted procedure is adopted and that creates a trust building even among operators and workers.

3. Since the "5S" methodology is the basis for any lean tool implementation and SMED demand high level of 5S implementation. In this way it provides platform for any industry to adopt other advance lean tools such as Kanban, Value Stream Mapping etc.

In this way SMED is used for not only setup time reduction but also having secondary benefits as mentioned above. SMED is more powerful if any industry and applies it with the integration of other lean tools.

Author details

Yash Dave
Department of Manufacturing Engineering, Government Autonomous Polytechnic College, Dhar, M.P., India

*Address all correspondence to: yashmegha@gmail.com

References

[1] Bajpai JD. SMED (single-minute exchange of die) methodology in garment manufacturing industry: Case study in reducing style change over time. InProceedings of the 5th International & 26th All India Manufacturing Technology, Design and Research Conference (AIMTDR 2014), Guwahati, India 2014 Dec (Vol. 1214).

[2] Raikar NA. Reduction in Setup Time by SMED Methodology: A Case Study. International Journal of Latest Trends in Engineering and Technology. 2015;5(4):56-60.

[3] Desai M, Rawani A. Productivity Improvement of Shaping Division of an automobile industry by using single minute exchange of die (SMED) methodology. ARPN Journal of Engineering and Applied Sciences. 2017;12(8).

[4] Ulutas B. An application of SMED Methodology, World Academy of Science. Eng Tech. 2011;55:100-3.

[5] Winatie A, Maharani BP, Riksa VH, Hasibuan S. Increasing Time Efficiency of Change over Process on Solid Product using SMED (Single Minute Exchange of Dies) Method in Pharmaceutical Industry.

Chapter 9

Lean Manufacturing as a Strategy for Continuous Improvement in Organizations

María Marcela Solís-Quinteros, Carolina Zayas-Márquez, Luis Alfredo Ávila-López and Teresa Carrillo-Gutirrez

Abstract

The implementation of lean manufacturing is one of the most discussed and studied topics in management; references are at the business, educational and public levels. However, the changes faced in the fourth industrial revolution generate challenges that will only encounter solution through innovative efforts and industrial improvements as well as a radical change in our way of interacting. In the current revolution, there are digital changes that cause ruptures in social, economic and political aspects, and the administrative process is part of it, this chapter proposes to analyze the implementation of lean manufacturing in the process of continuous improvement in business organizations through a literary review of the implementation of tools.

Keywords: continuous improvement, industrial revolution, digital, lean tools, high quality, globalization

1. Introduction

Despite being in many fields, organizations continue to be the spearhead in the innovation process. In the competition to produce high quality, low cost products, industries want to be more agile and faster. The environment of modernity forces companies to compete in markets outside their national borders, it is itself part of globalization, which as we know has played a very important role in the development of international finance. Multinational companies have managed to position themselves as one of the engines of the economy, coexisting in complex and heterogeneous environments.

In the search for quality, there is a fierce competition of globalization; organizations have invested in the development of knowledge and technology, improving already established processes. However, progress has not been homogeneous in all areas.

Lean manufacturing implementation is more important in some areas, as different tools are used to benefit the company and its employees. Some of the benefits include: reduction in production costs, reduction of inventories, reduction of delivery time (lead time), better quality, less labor, greater equipment efficiency, reduction of waste, overproduction, time of waiting (delays), transportation, inventories, movements, poor quality, among others.

2. Origins of lean manufacturing

Lean Manufacturing has its roots in the Toyota Production System originated at Toyota Motor Company by Taiichi Ohno and Shigeo Shingo [1]. It focuses on developing high quality, low cost products using less time, less space, fewer workers, and fewer tools.

After the First World War, Henry Ford and Alfred Sloan (General Motors) changed artisan manufacturing – used for centuries and directed by European companies– for mass manufacturing. Largely as a result, the United States soon dominated the world economy [2].

In 1950 Eiji Toyoda visited Ford's Rouge plant in Detroit for three months, an uncle had visited it in 1929. The Toyota Motor Company was founded in 1937. In 1950, after 13 years of work and effort they produced 2,685 automobiles, compared with the 7,000 they produced daily.

According to [3] Toyota's way of achieving a "lean" approach was to eliminate all waste, that is, activities that do not add value to the product from the customer's point of view. This allows reducing costs and increasing productivity. However, waste disposal is not enough because it requires a context and a culture, known and understood by all stakeholders (senior management, collaborators and suppliers). The Toyota way in a model is in a pyramid representing Toyota culture from top to bottom. This is the 4P model: 1) a philosophy of long-term thinking; 2) continuous process improvement to eliminate waste; 3) People and partners respect, challenge and grow; 4) Problem solving through continuous improvement and learning.

To locate the origin of the word lean, it is necessary to name JP Womack, and Daniel Jones, two researchers from the United States and from England, respectively, who are referents in this matter, through their work entitled "Lean Thinking" managed to concentrate the fundamentals of thought 'lean' and the concept of waste ('muda'). They also managed to perfect the concepts of value stream, flow, pull, among others.

Currently, a company that does not consider the implementation of this system is not in a position to compete in the world, since it will be absent from aspects such as the quality of products and services, reduction of operating costs and the subsequent increase in sales. However, to find the right tool it is necessary to carry out a detailed analysis of the company's conditions.

It can be said that this system places a company as a world-class one, since it carries out procedures that guarantee productivity, efficiency and the quality of products and services. In addition, several previous works confirm the benefits of implementing this system.

In the study [4] many factories reported positive results with the implementation of lean manufacturing. Some of these benefits are improved quality, high inventory turnover, and productivity increases. All these benefits lead to greater customer satisfaction and loyalty, and higher profits, in addition [5] comments that one of the most important criteria of this system is to get rid of unnecessary issues in manufacturing that do not add value to the product.

We observe in [6] the current situation in the implementation of lean practices in manufacturing plants in India and its impact on operational performance. The study was conducted by applying a questionnaire to 79 plants in different regions from India. Lean manufacturing is a multidimensional construct, finding that 80% of the plants have implemented various dimensions of lean manufacturing; such as focus on the needs of the customer, pulled production systems, reduction of exchange time of molds (SMED), total productive maintenance (TPM), relationship with suppliers, statistical control of processes and inclusive problem solving.

It was also found that productivity, quality at first intention, reduced delivery time and inventories, as well as occupied space, in summary, respondents stated that quality at first intention, reduction of time delivery and increased productivity are the top three drivers of lean adoption in that context.

Finally [2] finds that the constructs of lean, sustainable manufacturing and continuous improvement have a direct, relevant, positive and statistically significant impact on the dependent construct.

3. Lean manufacturing challenges

The fourth industrial revolution or revolution 4.0 is the product of the innovative efforts of the previous revolutions. Without a doubt, this phenomenon not only generated industrial improvements but also changes in the way of interacting with each other. Series production, electricity, transportation, are just a few examples of progress. In the current revolution, digital changes arise that cause ruptures in social, economic and political aspects, and in this sense, the administrative process is part of it.

In light of the technological developments associated with the revolution 4.0, the relationship between lean manufacturing and technology has once again become an area of research interest [7, 8].

The fourth industrial revolution promises to change the manufacturing landscape, and those who cannot take advantage of new technology-induced opportunities are destined to lag behind their competitors. An important area to investigate is the role that lean manufacturing will play in this new industrial age [7].

Lean manufacturing as a system is possible with various subsystems (tools), which are used to reduce and eliminate waste in companies. The right tools that help to eliminate all the waste and all the operations that do not add value to the product or to the processes, is the last aim of this system.

The challenges of lean manufacturing, especially in small and medium-sized companies have to influence the integration of the lean culture within their vision, develop leadership based on that culture. It is necessary to deploy that culture to everyone in the organization; involve the entire management team; have well-defined plans; know very well what processes contribute and which ones do not add value; and finally make this culture your day to day.

Below we give an account of the identification tools used in this system and we delve into the innovation of manufacturing, ending our analysis with the culture of quality.

The production systems related to a product or service must be analyzed under the approach of a value chain, which implies showing all the related activities to identify those activities that do not add value and that are classified as waste according to the manufacturing philosophy. Lean, which provides us with a series of tools that must be carefully selected to apply the most appropriate to eliminate waste and improve flow, as well as productivity in these production systems [9].

4. Importance of sorting waste according to the lean manufacturing philosophy

It is essential to know the types of waste that lean manufacturing proposes, also known as "Muda" due to its origin in Japan in the Toyota production system. Waste is anything that is not absolutely essential for production. Lean

Manufacturing identifies the following seven types of waste that must be eliminated in a production system [10].

Overproduction. Process items earlier or in greater quantity than required by the customer.

Wait time. Products or goods that are not being transported or in the manufacturing process are on hold.

Transport. The handling and transfer of materials or documents that do not add value are considered wasteful.

Inventory. Excessive storage of raw material, product in process and finished product.

Over-Processing. It is about adding unnecessary steps in work activities and not required by the client.

Movement. Any movement that the operator makes apart from generating added value to the product or service.

Defects in the process. Repetition or correction of processes also includes rework on non-conforming products or products returned by the customer.

Once the activities of the process where some type of waste is being generated have been classified, it is necessary that all the employees of the organization participate by undertaking specific actions for the elimination, for which the managers must promote an environment that promotes the generation of ideas and the continuous elimination of waste. In addition, it is important to train in the different tools, techniques and methodologies that lean manufacturing relies on and to establish a deadline to verify the results.

It is necessary to note that the full identification and elimination of waste has become a pillar for companies, when deciding absolutely on the use, investment and management of their resources, which is essential to find opportunities for improvement in the organization and achieve greater process efficiency [11].

5. Top lean manufacturing tools

The lean manufacturing tools that are mostly used in the industrial sector to increase productivity are the following: [12].

5S is a tool focused on working with effectiveness, organization and standardization. It is used to improve work conditions, through the organization, classification, cleanliness and order in the workplace. It seeks to establish a pleasant and high-performance work environment that allows the correct performance of daily operations [13].

According to [14], **Kaizen** is a Japanese word that means continuous improvement and presents a methodology to improve any production process through the implementation of different techniques, tools and methods, in addition to promoting a cultural change, teamwork, skills development and initiative to identify root causes of problems.

Poka Yoke seeks to prevent the production of defects caused by humans when they make mistakes in their workplace, blocking or stopping the process in case of failure [15].

Andón is the signal system that allows showing the state in which the machine is (running, stopped, stuck, etc.) [15].

SMED Single-Minute Exchange of Dies Set of techniques that seek to reduce machine setup times, Standardization through the installation of new mechanisms, templates and functional anchors, eliminates downtime adjustments [15].

TPM Total Productive Maintenance Set of techniques aimed at eliminating breakdowns through the participation and motivation of all employees, maximizing effectiveness and lengthening the life of the equipment and preventing losses in operations [16].

Kanban is a tool that helps to improve the production flow in a line through "labels", which serves as a "work order, informs in a timely manner what is going to be produced, in what quantity, by what means, and in what will be transported.

Jidoka Technique based on the incorporation of systems and devices that give machines the ability to detect failure in production.

Heijunka is key to achieving flow and leveling in production, the purpose is to level the production schedule by sequencing the orders according to a repetitive pattern that makes the average daily production similar between the different days of the week [16].

MUDA type	Lean Manufacturing Tool
Overproduction	• Kaizen (improvement event, work teams) • VSM (mapping of cycle times in production processes) • Heijunka (level production) • SMED (reduce setup times)
Timeouts	• Kaizen (application of the methodology for the proposal to reduce waiting times in the value chain) • VSM (identify and adjust lead times in production) • SMED (reduce waiting times in machine preparation) • Jidoka (automation with presence)
Transport	• Kaizen (continuous focus on improvement and improvement of logistics processes) • VSM (identification of transports in the different stages of the process)
Inventory	• Just in time (eliminate all inventory that represents waste in purchasing, manufacturing, distribution activities) • VSM (transfer of raw materials and finished product. From supplier to plant or from plant to customer. • Andon system (visual control of inventory levels) • Kaizen (kaizen event to improve inventory system) • Kanban (level the flow of materials)
Over-processing	• Kaizen (kaizen event to eliminate unnecessary steps in work activities) • VSM (process mapping to identify activities that do not add value from the customer's point of view)
Unnecessary movements	• 5 s (organize work so that unnecessary movement is minimized, ensures work areas are consistently clean and organized) • VSM (identify on the map the movements that are found and that do not add value to the client)
Defects	• Poka-Yoke (Anti-error system) • Jidoka (automation with people control) • Kaizen (continuous improvement proposal) • VSM (value chain mapping to identify defects) • Kanban, 5 s and Andon System (Visual control) • TPM (preventive maintenance plans)

Table 1.
Lean manufacturing tools for Mudas removal.

JIT Just in time. This system consists of getting the parts necessary for assembly to the line at the right time they are needed and in the right amount required.

VSM Value Stream Map is important to manage the entire value stream for each product, in order to understand where waste occurs, rather than looking at activities or processes in isolation. Therefore, a value flow map must be created where each of the actions required to design and manufacture a specific product is clearly identified.

Activities that create value from the customer's point of view; as well as those activities that do not create value from the customer's point of view, cannot be eliminated. They can be improved in order to reduce the time that they are given, and finally the activities that do not create value from the customer's point of view, and that are unnecessary, and these can be eliminated immediately [17].

The implementation of quality tools in a lean production system is strongly linked to common sense, and that is why adequate preparation in the organizational culture must be required, where both managers and employees have the commitment to change their traditional ways of thinking and working [18].

If there is not clear identification of the different quality tools that are applicable for the elimination of the different wastes, then no efforts aimed at improving the production processes will have the expected impact. That is why as a contribution to this work, **Table 1** shows the different changes and the lean tools that may be appropriate to solve these problems were identified.

6. Three IT (information technology) innovation trends impacting lean manufacturing

According to [19] one of the industries with the greatest global impact is the manufacturing industry, in which there is high competition worldwide. Therefore, he recommends that companies become aware of the transformation they must have in three areas: Internet of things (IoT), Smart factory and Personnel management.

6.1 Internet of Things (IoT)

Within the manufacturing sector, identification devices such as RFID (Radio Frequency Identification) or RTLS (Real Time Location System) are key to optimizing and planning process automation [19].

6.2 Smart factory

More and more companies are interested in adopting digital approaches in their processes to optimize costs and times. One of the great transformers in the sector is business software that integrates processes, since they help improve the profit margin or reduce machine downtime. An example is ERPs (enterprise resource planning system), which benefit the generation of proactive analysis, the reduction of errors or the monitoring of payments and collections [19].

In Mexico, 5S's application software was for MSMEs with the aim of being an engine of change towards continuous improvement [20]. Transforming the way of carrying out day-to-day operations in the MSME field is the innovative contribution that is intended to achieve it with the development of a 5S's application software. An improvement to the software is considered by adding four more S's, Shikari, Shitsukoku, Seishoo and Seido (Constancy, Commitment, Coordination and unification through standards); that is, implement up to 9S's in companies. Likewise, the 5 W (5 why), poka yoke and visual control tools can be developed in the same software.

The implementation of the 5S tool in all SMEs, aims to eliminate waste and ensure a clean and orderly work environment. The results of the application of this lean tool in SMEs are immediate, the result is a great visual impact, avoiding customer complaints, improving staff participation and improving process efficiency [21–23].

6.3 Personnel management

The manufacturing industry is one of the most questioned when it comes to staff turnover. The sector must be more sensitive to the importance of retaining the best-qualified workers because their potential generates greater competitiveness. It invests in the training, promotion and well-being of workers. It also detects areas of opportunity and improvement in the functions they perform within the organization [19].

According to [24] successful innovation requires a strategic and tactical deployment in the organization, integrating the commitment, resources, capabilities, and joint efforts of the different dependencies of the company, from senior management to operators. This includes engineering, research and development (R&D), production, marketing, merchandising, finance, human resources, and other functions.

In the context of lean manufacturing and its multiple tools, continuous improvement is immersed (**Table 2**). This is a key element in quality standards and processes. Currently in the quality culture of companies, the main standard is precisely continuous improvement, whose philosophy has permeated in various sectors around the world. A typical example is the Japanese companies, which lead the application of the tools of lean manufacturing, quality management and continuous improvement. This work methodology is about the Toyota Company, which is currently a

Tool	Basic purpose	Preliminary targeting	Method
Strategic tools	Strategic innovation tools are mechanisms that the company uses to analyze its internal resources and capabilities and external threats and opportunities.	Company strategy, useful for the entire organization. You get a SWOT of it.	A matrix is completed that gathers the characteristics of the organization and the behavior of the environment for the businesses developed.
Creativity	They are mechanisms that the company uses to promote creativity and promote the flow of ideas. Creativity is used either for the search for creative solutions or for the design of new products.	All the company.	Carry out a creativity session through brainstorming.
Technological surveillance	Know the main trends in the environment and technological leaders, in order to make decisions.	Company strategy, useful for the entire organization.	Databases or patents are analyzed and processed and in the end a technological map with different clusters is obtained.
Organizational capabilities analysis	Know how the organization's capacities are in innovation, external and internal, to take improvement measures.	Company strategy, useful for the entire organization.	Apply a questionnaire and in the end different graphs are obtained where the capacities are measured.

Table 2.
Tools related to innovation.

model to follow in lean manufacturing since it achieved a balance in the combination of strategy design and operational excellence.

Some of the tools mostly used in companies, aimed at the operational area are 5 s (Seiri, Seiton, Seiso, Seiketsu, Shitsuke), which is applied in companies basically to avoid waste, improve performance and efficiency and thereby improve productivity in the workplace. SMED (Single Minute Exchange of Die), its usefulness in the processes is to reduce the change time and increase the reliability of the change process itself. TPM (Total Productive Maintenance), focuses on managing the maintenance of the company in order to maintain zero failures, here all personnel and all phases of the production process are involved. Kanban serves to identify the material requirements made in the production process [25].

JIT (Just In Time) tools, which originated with the purpose of eliminating inventory, are also widely used, in addition to other diagnostic and monitoring tools, automatic stops, takt time, level production etc. These tools originally created in the Japanese automotive industry, with a notable deployment to other sectors of the maquiladora industry, but especially in other areas and in small or medium-sized companies around the world.

In summary, lean manufacturing has triggered a series of steps that allow a true adherence to excellence, among these steps is continuous improvement, the objective of which is to achieve standards and good quality management. In this sense [26] mention that the starting point for companies is to have an organization focused on the client, understand current and future needs and try to satisfy them. Another essential aspect is having excellent leadership through which the staff is involved and thereby enhances the participation of different levels and areas of the company, which translates into a high level of commitment.

In the culture of quality, emphasis is on the focus on systematized processes directed towards quality management. These aspects result in a constant evolution towards quality and lead to efficient decision making that benefits both the forward processes, that is, with the client; as backwards, that is to say with the providers; thus achieving a convergence in all the processes involved.

Historically, there have been great companies and brands that have made revolutionary contributions to quality. Proof of this is undoubtedly Ford Motor Company, American Telephone & Telegraph and Western Electric, whose processes initiated the implementation of quality criteria in various areas, not only in production. On this same subject, the participation of Japanese companies has been more than relevant, since in addition to Toyota, "Company Wide Quality Control" was created, which involves all the resources and processes existing in the company, thus achieving what is called "total quality" [27].

On the other hand, in Europe there is the "European Foundation For Quality Management" that synthesizes its quality model in the satisfaction of customers and employees through leadership that promotes policies and strategies of the organization. Along with an adequate implementation of the resources and above all a perfect management of the most important processes in the organization to obtain excellence in results [28].

As for the Latin American region, there are also records of the implementation of lean manufacturing tools as a strategy for achieving quality. In the study [29] analyzed Mexican and Colombian companies, and applied an instrument where they were asked, among other things, how often they use tools for quality management. The responses showed a clear trend towards use in managerial positions and from middle to senior management onwards. What is interesting about this analysis is that it focused on small and medium-sized companies, which makes it clear that the culture of quality has permeated not only in various areas, but also in companies from different sectors.

In summary, lean manufacturing provides great benefits to companies by eliminating waste, processes that do not generate value and that make administration bureaucratic and expensive. The new projects seek to be born in the context of continuous improvement and quality culture, and on the other hand, existing companies make great efforts to incorporate these work methodologies, seeking as their main purpose the reduction of operating costs.

It is clear that the adoption of lean manufacturing is on the way towards efficiency in the processes of the main line of business of companies. It is evidently focused on permeating the organizational culture in order to transform the competitiveness of the company; and thus be more attractive to shareholders, employees, suppliers and customers. Every time of a demand of better conditions not only in the products, but also in the prices, response times of the company, the commitment that it shows on social responsibility and environmental responsibility; and in general in the image that is projected towards the market [30].

For companies to have the ability to respond to the new and increasingly demanding market requirements, the decision to adopt the culture of quality is necessary first. It corresponds to managers or owners of the companies, probably encouraged by the workers themselves who are in the operational area and know and experience the problems that carry out their work in a daily way.

A part of special importance in such a large organizational change is to establish, provide and promote the appropriate conditions for the development of this type of project. Carrying out an organizational transformation entails re-educating all the parties involved. From suppliers, at the time they are requested to implement computer systems that make deliveries just in time and without defects, through production that must focus on zero errors, zero waste and zero delays, encouraging real production cost calculations and exact, documentation, improvement and control of each of the processes, as well as efficient and effective management of resources. All the departments that together make up the company must also be involved, that is; continuous improvement and lean manufacturing must permeate the entire organization.

These business transformations must emerge from the organizational culture itself, it does not mean a change of personnel to generate new uses and customs, but an evolution in the way in which business processes are carried out, trying to reduce the maximum movements and the number of processes and transactions carried out to achieve production.

Among the main elements of the culture of quality, it is suggested to bear in mind that perfection does not exist; however, processes can work in a harmonious, uninterrupted way and with immediate error detection to avoid waste and rework. It is important to have an open mind that socializes the information openly, and above all to talk and share problems as well as opportunities for improvement, for example holding kaizen events where problems are analyzed to obtain holistic resolutions that involve the entire company.

Another essential aspect in the culture of continuous improvement is to instruct workers to maintain a dynamic based on learning, which is also systematized and is not linked especially to one person, but to the work team in general. This is important given that when a problem is studied and all parties participate, each department exposes its intervention in the analyzed process and the way in which they can intercede to improve and at the same time reduce the amount of processes, costs, and waste that were generated.

Once this open culture is established, the next level is to encourage accompaniment to support people in the development of their processes, and remain receptive to the emergence of new ways of doing things, accept, adopt and adapt changes that generate added value, lower costs and provide greater agility in deliveries. It should not be forgotten that in lean manufacturing there are four important parts that must be coordinated to achieve perfection in the systems. According to [31] the starting point is the design and engineering of the product, the supply chain, the demand and the customer.

The design is based on teamwork made up of members who come from different departments, with experienced leaders who encourage effective and respectful communication, where all participants must express acceptance of the decisions made, with the aim of that differences and conflicts are exposed at the beginning of the projects and not afterwards.

On the other hand, companies require excellent coordination in the supply chain to have access to good quality materials, with fair prices and in a timely manner. In the lean philosophy, functional levels are proposed and each of them has its own responsibilities. The first level is the integral part of the design and development of a new product. The second level is where the necessary parts are supplied. It should be borne in mind that lean manufacturing considers the customer first as the guiding principle of what to do. Therefore, companies must adapt to the market and the constant transformation of demand.

It is worth mentioning that every day the market and with it the demand, tastes and preferences of consumers continue to globalize. This fact has led small companies to incorporate lean manufacturing practices since to be part of the global value chains of the production systems of large companies (transnational, multinational, international and global), the main requirement is the management of quality demonstrated from production under standardization.

In this increasing integration, it is impossible to go back and regionalize production since supply chains are interconnected globally, which means that production is also global. In this sense, it is urgent, especially for micro, small and medium-sized companies, to join what has also been called the Toyota house, whose main objective is excellence in operations [32].

The key piece of this transformation is undoubtedly a human talent that enjoys commitment, academic training, professional experience, communication and leadership skills and with sufficient motivation to direct their efforts towards the realization of stable and standardized processes that result in a level production, always working under the motto of continuous improvement.

7. Conclusions

This chapter reviewed the application and integration of lean manufacturing tools in a current perspective, emphasizing the continuous improvement, detailed analysis of the current state of the process, and identification of a reliable starting point; in addition to the commitment of employees and management in the stages of planning, monitoring and taking actions.

Finally, it becomes clear that lean manufacturing increases quality and productivity in companies by reducing waste and therefore production costs; adapting favorably to the different innovation systems that are required today; and it enables the establishment of a continuous improvement work methodology that invites the constant review of processes and consequently the culture of change and quality within companies.

Author details

María Marcela Solís-Quinteros, Carolina Zayas-Márquez, Luis Alfredo Ávila-López*
and Teresa Carrillo-Gutirrez
Universidad Autónoma de Baja California, Tijuana, Mexico

*Address all correspondence to: avila.luis@uabc.edu.mx

References

[1] Gadre, A., Cudney, E., & Corns, S. Model development of a virtual learning environment to enhance lean education. Procedia Computer Science; 2011. 6, 100-105.

[2] Padilla, L: Lean manufacturing manufactura esbelta/ágil. Revista Electrónica Ingeniería Primero; 2010. ISSN, 2076(3166), 91-98.

[3] Alves, A. C., Kahlen, F. J., Flumerfelt, S., & Manalang, A. B. S: The lean production multidisciplinary: from operations to education; 2014

[4] Elbadawi, I., McWilliams, D. L., & Tetteh, E. G. Enhancing lean manufacturing learning experience through hands-on simulation. Simulation & gaming; 2010. 41(4), 537-552.

[5] Mady, S. A., Arqawi, S. M., Al Shobaki, M. J., & Abu-Naser, S. S. Lean Manufacturing Dimensions and Its Relationship in Promoting the Improvement of Production Processes in Industrial Companies. International Journal on Emerging Technologies; 2020. 11(3): 881-896(2020)

[6] Ghosh, M. Lean Manufacturing performance in indian manufacturing plants, Journal of Manufacturing Technology Management; 2013. 24(1) 113-122

[7] Buer, S. V., Semini, M., Strandhagen, J. O., & Sgarbossa, F.: The complementary effect of lean manufacturing and digitalisation on operational performance. International Journal of Production Research; 2020. 1-17.

[8] Gupta, S. , S. Modgil , and A. Gunasekaran. Big Data in Lean six Sigma: A Review and Further Research Directions. International Journal of Production Research; 2020. 58 (3): 947-969.

[9] Wilches, M. J., Cabarcas, J. C., Lucuara, J. & Gonzalez, R. Aplicación de herramientas de manufactura esbelta para el mejoramiento de la cadena de valor de una línea de producción de sillas para oficina. Revista Dimensión Empresarial;2013. vol. 11, Núm. 1, pp. 126-136.

[10] Chase, R. B. & Jacobs, F. R. Administración de operaciones: producción y cadena de suministros. McGraw-Hill; 2019.

[11] Corredor, I. Sin identificación de los 7 desperdicios no hay Lean (Tesis de maestría). UNAM; 2015.

[12] Favela-Herrera, M., Escobedo-Portillo, M., Romero-López, R. & Hernández-Gómez, J. Herramientas de manufactura esbelta que inciden en la productividad de una organización: modelo conceptual propuesto1. Revista Lasallista de Investigación; 2019. 16(1), 115-133.

[13] Vázquez, J. Indicadores de evaluación de la implementación del lean manufacturing en la industria (Tesis de maestría). Universidad de Valladolid; 2013.

[14] Rieger, T. Desarrollo de un método visual para el análisis y la evaluación del nivel de productividad en la fabricación industrial basado en la variación del flujo de valor (Tesis doctoral). Universidad Nacional de Educación a Distancia, Madrid; 2011.

[15] Hernández Matías, J. C., & Vizán Idoipe, A. Lean Manufacturing. Concepto, técnicas e implantación (p. 174). MADRID: Escuela de Organización Industrial; 2013.

[16] Villaseñor, A. & Galindo, E. Manual de Lean Manufacturing. Guía básica. Editorial Limusa; 2009.

[17] Tapia Coronado, Jessica, Escobedo Portillo, Teresa, Barrón López, Enrique, Martínez Moreno, Guillermina, & Estebané Ortega, Virginia. Marco de Referencia de la Aplicación de Manufactura Esbelta en la Industria. Ciencia & trabajo; 2017. 19(60), 171-178. https://dx.doi.org/10.4067/S0718-24492017000300171

[18] Cabarcas Reyes, J., Wilches-Arango, M., Forero Chaves, A., & Molina Sanmiguel, S.: Análisis y mejoramiento de la cadena de valor de la línea de producción de láminas de una empresa del sector metalmecánico mediante la aplicación de herramientas de manufactura Lean. INGE CUC; 2011. 7(1), 27-42

[19] Becerra J. Tres tendencias de innovación TI en el sector de manufactura. CIO México; 2019.

[20] Romero-Cruz C.; López-Muñoz J.; Méndez-Hernández J. y Pintor-Tuxpan A. Software para implementación de 5S's en Mipymes y su relación con la mejora continua y la competitividad. Revista de Negocios & PyMes. 2016; 2(5): 45-53.

[21] Manzano Ramírez, M. y Gisbert Soler, V. Lean Manufacturing : implantación 5S. 3C Tecnología de innovación aplicadas a la pyme; 2016. 5(4), 16-26. DOI: http://dx.doi.org/10.17993/3ctecno.2016.v5n4e20.16-26/.

[22] Piñero, Edgar Alexander, y Vivas Vivas, Fe Esperanza, y Flores de Valga, Lilian Kaviria, Programa 5S's para el mejoramiento continuo de la calidad y la productividad en los puestos de trabajo. Ingeniería Industrial. Actualidad y Nuevas Tendencias, vol. VI, no. 20; 2018. 99-110.

[23] Jaen, F., Villanueva, V., y Novillo , E. Análisis y propuesta de mejora de procesos aplicando 5s en una empresa de mantenimiento. Caso Ecuaclima. 593 Digital Publisher CEIT; 2020. 5(3), 27-37. https://doi.org/10.33386/593dp.2020.3.207

[24] Arrieta Canchila, Katty Milena: Diseño de una metodología que relaciona las técnicas de manufactura esbelta con la gestión de la innovación: una investigación en el sector de confecciones de Cartagena (Colombia). Universidad & Empresa; 2015. 17(28)127-145.

[25] Ehni, M., & Kersten, W. Toyota Kata: Empowering Employees for Target-Oriented Improvement-A Best Practice Approach. In Innovations and Strategies for Logistics and Supply Chains: Technologies, Business Models and Risk Management. Proceedings of the Hamburg International Conference of Logistics (HICL); 2015. 175-200

[26] García, M., Quispe, C., & Ráez, L. Mejora continua de la calidad en los procesos. Industrial data; 2003. 6(1), 89-94.

[27] Arbós, L. C., & Babón, J. G. Gestión integral de la calidad: implantación, control y certificación. Profit editorial; 2017.

[28] Chacon, j. & Rugel, S.: Artículo de Revisión. Teorías, Modelos y Sistemas de Gestión de Calidad. Revista Espacios; 2018. 39(50).

[29] Cruz Páez, F. O., Ibarra Díaz, C., Rueda Pineda, D. J., & Olivares Vera, D. A. Análisis exploratorio sobre la apreciación de características predominantes en empresas medianas de México y Colombia en temas de calidad, competitividad, innovación social y productiva. Tendencias; 2020. 21(1), 130-156.

[30] Vargas-Hernández, J. G., Muratalla-Bautista, G., & Jiménez-Castillo, M. Lean Manufacturing una herramienta de mejora de un sistema de producción?. Ingeniería Industrial. Actualidad y Nuevas Tendencias; 2016. (17), 153-174.

[31] Tejeda, A. Mejoras de lean manufacturing en los sistemas productivos. Revista ciencia y sociedad; 2011. 36(2) 276-310.

[32] Bautista, S. J., & Peralta, C. M. R.: La inclusión de las PyMEs en la Cadena de valor de la Industria Automotriz en México en el marco del Tratado Trans-Pacífico (TTP). Economía Informa; 2017. 403, 46-65.

Chapter 10

Development of Integrated Lean Six Sigma-Baldrige Framework for Manufacturing Waste Minimization: A Case of NAS Foods Plc

Kassu Jilcha Sileyew and Selamawit Gebreyohanis

Abstract

The aim of this study objective is to develop an integrated constant quality improvement model so as to minimize unwanted biscuit processing industry wastes. The method used was lean- six- sigma elements to define measure and improve unwanted process company wastes. In other word, Baldrige with six-sigma were created to define, measure and improve management perspectives. The tasks were integrated using both quantitative and qualitative analyzing tools implementing mixed strategies. The result was improved by using FMEA analysis was carried out at each stage of the existing process used to determine the failure of the process and to analyses and improve the production quality. The SPSS software was also used. In the finding section, the correlation and regression analysis has shown that there is strong relationship between each variance. There are different wastes that identified in six sigma (DMAIC) on NAS food Plc as a result; the value of waste ratio indicated is 36.7%. This show non-lean of the food industry is practiced. The defect of the company also calculated and defect per million are 67,308. This shows that the biscuit production has a production capability with a failure of 67,308 every 1000,000 productions it high failure rate. The contribution of the paper has indicated that there are limited studies were conducted so far to implement waste minimization tools like six-sigma, lean and MBNQA framework approach integration for food processing industry.

Keywords: waste, lean, DAMIC, CQI, MBNQA

1. Introduction

Companies are more competitive in current globalization and every detail is important for the business that wants to improve its competiveness. Since it is not a surprise, the best continuous improvement strategies that could be developed for manufacturing industries provide to improve efficiency and effectiveness of whole systems. Though, automating production and improving process efficiency are two major objectives of the food industry worldwide. Since, implementing advanced continues an improvement strategy in the modern manufacturing provides to improve the performance of the process, reduce waste and ensure on time delivery of the sectors [1]. Because of the continuous improvement is a management philosophy that approaches to tackle the challenge of product and increase process improvement [2]. Thus, from numerous tools firms were invested to implement lean, six-sigma, TQM and JIT strategies in their business process so as to enhance performance and compete at global levels [3]. Though the lean principle provides to identifying and eliminating non value add (wastes) through continuous improvement tools, flowing the product at the pull of the customer in pursuit of perfection [4]. Since due to comprise on quality development, process focus, continuous improvement and value stream management and worker empowerment future several companies invest in the implementation lean principles [3]. While six sigma is a really effective tool for systematically attacking the highest priority production and support functional problems within an organization [4]. Hence, six sigma is statistical measure of defect rate within a system and the practice requires the knowledge of basic and advanced statistical tools so as to reduce defects and variations within a work process in firm. But the main limitations are six sigma not effective for every problem reduction in the firms [5]. Baldrige national quality award (MBNQA) for helping excellence ever, improving value to marketplace success, improvement of overall organizational effectiveness and capabilities, organizational and personal learning. Though from the above observe that a single improvement tools lacks to coherence to tackle the whole problems of the manufacturing firms. Single tools have good potential to tackle specific problems. Even though, an integrated continuous improvement strategies and methods have great power to improve the performance and the competitiveness of the firms [6]. As long as fewer studies were investigate to implements the integration of lean with six sigma, lean with TQM strategies in the firm so as to improve the performance and global Competitiveness [4]. Therefore integrating three or more continuous improvement strategies provides to enhance, quality of products, optimum product cost, timely delivery of products,the flexibility of their internal business process, reduction of wastes and customer satisfactions [3]. Thus, this study aims to develop an integrated frame work from lean, six sigma and the Malcolm Baldrige Criteria so as to improve the production performance, reduce waste and improve competitiveness of NAS Foods Plc.

2. Problem statements

Nowadays food industry makes a significant contribution to national economy in many developing countries. Here in Ethiopia the economy depend on agricultural and which accounts 43.2% of gross domestic product (GDP) [7]. Also there is less contribution in an improvement of biscuit industry analyzed and waste minimization. There is different waste that are identified in six sigma (DMAIC) on NAS food Plc. Currently the company only uses 30% of its capacity and produce 3000 carton per day via 349 workers but it can produce 10,000 cartons with 1000 workers. As a result the value of waste ratio is 36.7%, this show non lean and can be categorized as a traditional company [8]. The defect of the company also calculated and defect per million are 67,308. This shows that the biscuit production of has a production capability with a failure of 67,308 every 1000,000 productions or equivalent to 6.73% loss and this indicates the production process still has a high failure rate. The other problem here is there are limited studies were conducted so far to implement waste minimization tools like six sigma, lean and MBNQA approach on Ethiopian food processing industries [9].

3. Objective

The main objective of this study is to develop an integrated continuous quality improvement model so as to minimize waste of biscuit manufacturing.

- To identify the gaps and strength of previous researches on related to continuous improvement strategies to minimize waste.
- To investigate the challenges and potentials of integrating continuous improvement strategies in the context of biscuit manufacturing.
- To develop an integrated continuous quality improvement models to reducing waste and enhancing competitiveness of NAS food Plc.

4. Literature review

The purpose of this literature review is to provide a background on challenges of waste, waste controlling mechanism and applicability of waste controlling strategies across a variety of industries. Since this understanding will help determine which waste controlling strategies and principles are appropriate for implementation within the food manufacturing industry, and detail topics discuss in this section are for history of waste minimization in manufacturing industry, lean manufacturing philosophy, six sigma and Malcolm Baldrige National Quality Award (MBNQA) principle, tools and technique in manufacturing industries as whole as well as a detailed summary of the literature concerning improvement and competitiveness problem of Ethiopian food processing industry. Though the principal sources of this information included company reports, published literature in textbooks and journals were incorporated.

4.1 Wastes in the food processing sector

Food loss should mean the decrease in edible food mass throughout the food chain. Food losses take place in production, postharvest and processing stages in the food supply chains. Since a division is to be made whether the loss of resources happens in the early stages of the food supply chains (FSC) or the resource was wasted by the action of the retail sector or consumers. Though in the first case, the problems can call about food losses, while in the latter case about food waste. The food losses can be avoided by a correct action, e.g. by maintaining the cold supply chain or ensuring correct storage conditions for products. Since the food loss also occurs if the product that was originally intended for human consumption is recovered in the form of feed, fertilizer or energy [10, 11]. Besides the waste and food waste is more comprehensive and it includes all resources that are lost in the different sectors of the food supply chain, and will include also those parts that were originally not intended for human consumption.

4.2 Waste minimization strategies and mechanisms

The waste is an important issue that should be treated in such a way that the benefits achieved from that will be in both environmental and social aspects. As in manufacturing industries, waste management is key issues to greater opportunities for waste recovery and diversion of waste from landfill, and services suitable to businesses. Though according to [10] shows that waste management is favoring reuse and recycling before land filling, with the main idea of environmental sustainable society.

4.3 Continuous improvement strategies

Quality is becoming an increasingly important subject in organizations. It is central matter to develop a sustained resource management technique. Therefore, logistics emerges as an activity that allows the achievement of a prodigious efficiency and economic welfares and in long term it is to obtain competitive advantages of the country. However, organizations have improvement choices with time depending up on the way they track to meet their strategic and operational objectives, they can watch to its economic welfares. Science raises continuous improvement as the competence of frequent processes and schemes and is closely integrated with means of waste elimination. The customers focus programs such as total quality management, supply chain management, just-in-time and kaizen also need to be understood how to assess quality of products and systems through use of a variety of quality control tools. It is also important to understand how to interpret findings and how to correct problems. There are countless specific tools and techniques that help the industry process improvement and enhancement. Any organization needs to focus on quality of products, optimum product cost, timely delivery of products and the flexibility of their internal business process to adjust to quick fluctuations. Consequently, continuous and constant improvement tools like Kaizen, Six Sigma, ISO-9000, Lean thinking, TQM, SCM and JIT system are not new to the present situation of industrialization for achieving the above concepts. This method is not only limited to any particular industry but also to large potential benefit. It has a widespread application throughout the whole industries. Many industries have adopted it and others are going to implement for their endurance in the fast competition at each stage in each area.

4.4 The concepts of lean manufacturing

Historically the lean production system is the world famous production system developed and practiced by Toyota Company for a long time [12, 13]. Though the basic ideas behind the lean manufacturing system are waste elimination, cost reduction and employee empowerment. Since this concepts leads to maximize customer value while minimizing all the wastes that come with that significance. Nevertheless, lean processes can make jobs highly repetitive while eliminating critical rest time for employees [14]. Lean is a philosophy that aims to maintain smooth production flow by continuously identifying and eliminating waste resulting in increasing value of activities in the production process [15].

4.5 Waste reduction and waste removal

Through Lean Main goal of lean thinking is to reduce and remove waste. Since lean strategy is a potential tool to minimize and then remove waste so as to achieve sustainable development of manufacturing firms. Even though, the lean strategy reduction is one of the main functions of Lean Manufacturing implementation plan [15]. Since all the form of waste i.e. overproduction, defect, transportation, work in progress inventory, over processing, waiting and motion are reduced with Lean manufacturing implementation (**Figure 1**).

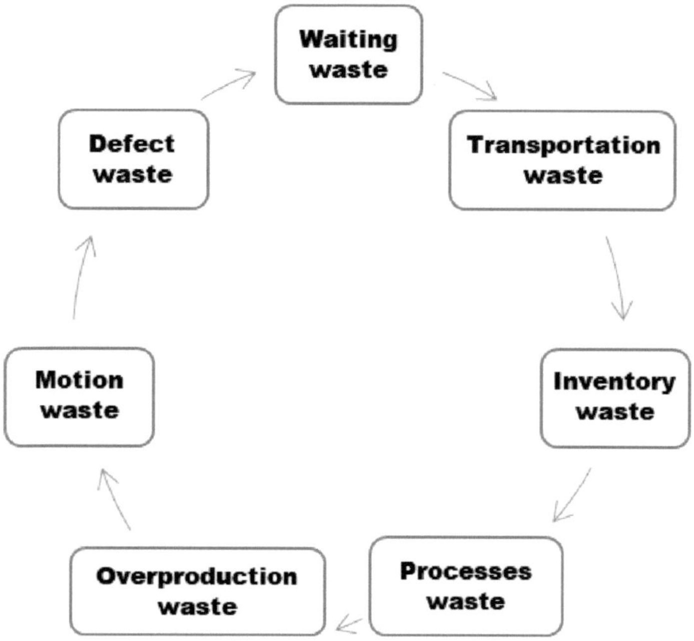

Figure 1.
The seven wastes that controlled quality.

4.6 Roles of six- sigma on improvement

Numerous companies use Six Sigma practice to achieve competitiveness of their business. Six sigma methodologies are cast-off to improve the excellence of the product and process dramatically. Sigma, σ, is a letter employed from the Greek alphabet to measure the process variability and the sigma level measured to determine the performance of the business processes [15]. The six sigma methodology was introduced by Womack et al. [16] and resulted in the accomplishment of business quality in Motorola. According to the study by Snee [17] states that six sigma concept was constructed by Bill Smith, then an engineer at Motorola who wins the 1988 Baldrige National Quality Award. The deployment of six sigma concept is led by Allied-Signal and General Electric (GE). Six-sigma is described as an improvement programmer for reducing variation. It focuses on continuous and breakthrough improvements. In research the two major improvement methodologies in six- sigma has been considered. These are already existing processes and new processes. The first methodology used to improve an existing process and can be divided into five phases [15]. These are:

- **Define:** which process that needs improvement. Define the most suitable team members to work with the improvement. Define the customers of the process, their needs and requirements, and create a map of the process that should be improved.
- **Measure:** Identify the key factors that have the most influence on the process, and decide upon how to measure them.
- **Analyze:** Analyze the factors that need improvements.

- **Improve:** Design and implement the most effective solution. Cost–benefit analyses should be used to identify the best solution.

- **Control:** Verify if the implementation was successful and ensure that the improvement sustains over time. As well the second methodology is often used when the existing processes do not satisfy the customers or are not able to achieve strategic business objectives, see [18]. This methodology can also be divided into five phases; define measure, analyze, design, verify, according to [15]. In summary, the two different methodologies have obvious similarities [19]. **Table 1** indicates the integration of the tools.

	Define	Measure	Analyze	Improve	Control
Lean tools	Value Mapping Project Charter	Process Mapping Cause and Effect Matrix	FMEA Bottleneck Analysis	Production Smoothing Kaizen Events	Standard Work 5S Poka-Yoke

Table 1.
Integration of lean tools in the DMAIC framework [3].

Program	Six Sigma	Lean thinking	MBNQA
Theory	Reduce variation	Remove waste	Quality awareness
Application and Principles	• Define • Measure • Analyze • Improve. • Control	• Identify value • Identify value stream. • Flow. • Pull, Perfection.	Visionary Leadership, Customer-Driven, Excellence, Organizational and Personal Learning, Valuing Employees &Partners, Agility, Focus on the Future, and analysis and knowledge transfer and organization performance
Focus	Problem focused	Flow focused	Quality focused
Criticisms	System interaction not considered. Processes improved independently.	Statistical or system analysis not valued	Unfairness, superficiality and publicity the inherent value of the continuously improving award program far outweighs its limitations.
Tools	Flow chart, control chart, graphical chart	5S,VSM	Questioner, ABC
Type of Continuous Quality Improvement (CQI) Initiative	• Emphasis processes & outcomes • Greatest for processes plagued by wide variability —logging of pharmaceuticals, standardizing referral processes, etc. • a heavily quantitative approach to CQI	• Emphasis on process. • Simplifies overcomplicated processes and considers interdependencies. • Best for known problems with known system change solution. • Integrated throughout the organization. • Ideal for large complex health care	• Emphasis on structure and outcomes. • Best for practice-wide problem assessment and goal setting. • A broad, holistic approach to CQI initiated at strategic times. • Ideal for practices that want to establish a new

Program	Six Sigma	Lean thinking	MBNQA
	• Adapted for targeted changes to specific processes. • Combined with Lean when the focus is on efficiency and quality. • Ideal for practices that want to rigorously quantify improvements in safety, quality, and cost effectiveness	organizations and practice networks that want to standardize operations across multiple units	CQI system or overhaul an existing one.

Table 2.
Comparison of improvement programs [22, 23].

4.7 Malcolm Baldrige national quality award

The initiative taken to improve quality management practices and the competitiveness of U.S. firms was signed by President Ronald Reagan on Malcolm Baldrige National Quality Improvement Act in 1987. The Malcolm Baldrige National Quality Award (MBNQA) was created to promote quality awareness, identify the requirements for quality excellence and share information about successful quality strategies and benefits [20]. The Baldrige core values and concepts includes visionary leadership management for innovation, customer-driven, excellence, management by fact, organizational and personal learning, social responsibility, valuing employees and partners, focus on results and creating value, agility systems perspective and focus on the future are the common values [21].

Table 2 indicates the comparison of the different programs and tools. It indicates the difference and similarities the tools have so that they lead to develop the ingrate once. **Figure 2** indicates the summery of literature review with the major areas considered during the study.

4.8 Literature gap

To get enough information about the topic raised so many literatures are reviewed from different sources, among these journal articles, reports, and unpublished master thesis is the main one. During literature survey recent

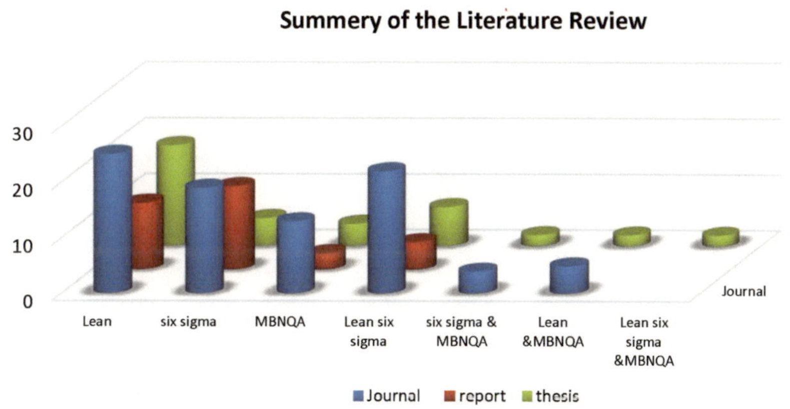

Figure 2.
Summary of the literature review (authors).

documents concerning waste minimization tool lean, six sigma and Malcolm baldrige quality award are collected from different sources then each document critically examined in order to filter gaps below.

- [9, 19, 23–29] this papers integration lean six sigma with different tools but not integration awards or quality perspective.
- [21, 30] these papers see the alignment and reviewed of Lean Six sigma and Baldrige but not integrate all and not much has been found in Ethiopian context.
- The organizations are not able to reap out the benefits of Lean Six sigma, Baldrige and other advance tools practices due to lack of awareness.
- Integrated Lean, Six sigma and Baldrige approach in Ethiopia Industries is not explored and not much has been found in food industry.

5. Research methodology

This study was conducted based on both secondary and primary data collected from the primary sources and ordinary data. Preliminary literature review and existing company condition was scanned to formulate the problems and objectives of the study. The data collection process considered defines measure and analyses the data sources. The process set improvement model and then control the research process. The research draws the conclusion of the resulting analysis with.

As shown in **Figure 3**, the research process start at problem formulation and arrives at conclusion and recommendation. The study has been conducted by considering preliminary literature review to develop objectives and problem statement. The study was conducted by considering literature review from different know sources and databases. The literature was reviewed from databases like Scopus

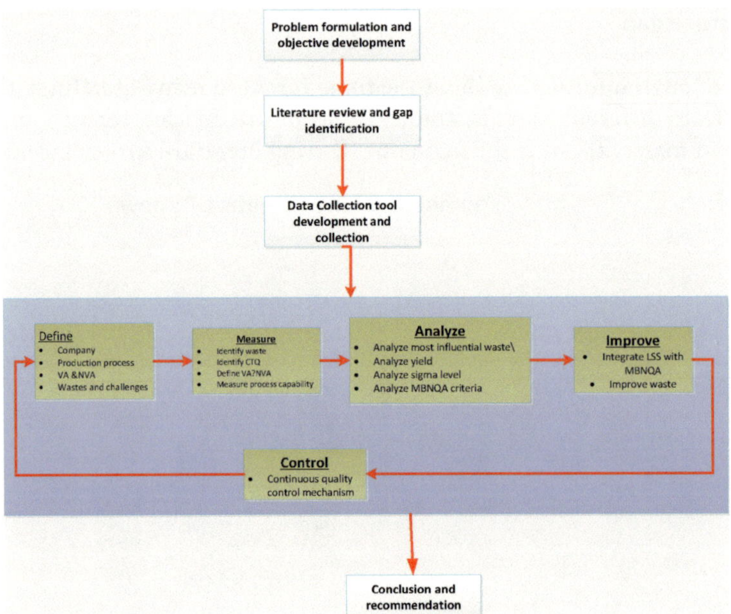

Figure 3.
A methodology framework.

indexed, web of science listed journals, PUBMED, MEDIN, research gates and DOJ indexed journals. After analyzing and screening literatures, the study found gaps from literature that helped to know the research focus area. Based on the literature review method of data collection and sources were identified. The data was collected through questionnaires, interviews, and field observation. Data sources were used from primary data which was collected by physical field observation, interviews, questioners and company reports from NAS Foods Plc. Responsible and targeted groups were considered under survey on this study. Interview of top managers were made containing 14 interview questions and answer by 1 management and 2 supervisor of the NAS Foods Plc. during the field visiting.

Questioners for employees also conducted to collect data from employees by using questioner to find detail of the problems that NAS Foods Plc. currently facing. The other data source was secondary data which was used to meet the research objectives, reviewing the existing research work of 48 journals, government reports, some reference books & paper related to lean thinking, six sigma, Malcolm Baldrige national quality award programs, strategies, role impacts on manufacturing and food processing industries. The key challenges, potentials and strategies to integrating continuous improvement tolls also considered. The Early search results show that a total of 400 article reports and thesis were found from various textbooks, academic and professional journals. Then read and sort for relevance to the continuous improvement and waste minimization strategy and tools and for their integration. The article would be assessed of methodology, method of measurement and finding results. Finally 48 article, reports and thesis selected are important and related to this study. The research methodology used the continuous quality improvement tool integration to reach its conclusion (Refer to **Figure 3**).

6. Result and discussion

6.1 Quantitative result

6.1.1 Bivariate correlation analysis

Correlation analysis is used to quantify the association between two continuous variables (between an independent and a dependent variable or between two independent variables. Pearson (r) correlation is the most widely used correlation statistic to measure the degree of the relationship between linearly related variables. The correlation value $r = -1$ indicated that strong negative correlation existence, $r = -0.5$ negative correlation, $r = 0$ with no correlation, $r = +0.5$ with strong correlation and $r = +1$ is the strong positive correlation (**Figure 4**).

This study showed the respondent result from questioner and it has 4 sections and 25 questions so in order to see the correlations of all indicators it is preferable to make analysis using SPSS. Based on the above principle the study develop the relationships between the waste measurement variance as we see in the next (**Table 3**) in the SPSS output Pearson correlation r (value of statistical test) should close to +1 and the sig (2-tailed) or p-value is less than 0.05 for strong relation. As we see the table below there is strong relation in each relation.

6.1.2 Analysis of awareness of waste measurement

Waste measures are included seven perspectives in lean typical. In this study it identified each because waste issue is different from process angle and to see further correlation between each viewpoint. Waste minimization is the basic for any

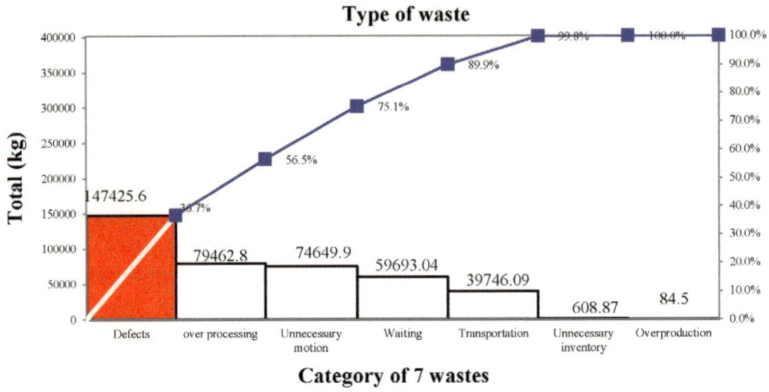

Figure 4.
Pareto diagrams of seven wastes)

Factors	Excessive transport	Unnecessary inventory	Unnecessary motion	Inappropriate processing	Overproduction	Waiting time	Defect
Excessive transport	1	0.749**	0.662**	0.579**	0.696**	0.630**	0.680**
Unnecessary inventory		1	0.898**	0.692**	0.661**	0.920**	0.779**
Unnecessary motion			1	0.882**	0.814**	0.899**	0.848**
Inappropriate processing				1	0.906**	0.762**	0.903**
Overproduction					1	0.659**	0.894**
Waiting time						1	0.828**
Defect							1

***Correlation is significant at 0.01 level with Person (2-tailed) and list wise N = 100.*

Table 3.
Bivariate correlation in between waste measurement parameter.

organization, in this study waste minimization of NAS food plc. The awareness of each respondent comprised (**Table 3**).

When it has been seen the relationship of each variable in waste perspective have strong relation with significance level of 0.01 and the causal Pearson Correlation of variable of excessive transport vs. inappropriate process their value is 0. 579 which show moderate positive relationship and the highest Pearson correlation in waste perspective is between unnecessary inventories vs. waiting their value is 0.920 it mean that waiting in NAS food is highest factor in unnecessary inventory analysis. There is also highest Pearson correlation that the value is greater than 0.9 between inappropriate processing vs. over production and inappropriate process vs. Defect.

6.2 Analysis of Malcolm criteria measurement

Malcolm measures are included seven criteria in management perspective. In this study it identified each because Excellence issue is diverse from management angle and to see further correlation between each viewpoint. Quality improvement is the basic for any organization (**Table 4**).

When we see the above relationship of each variables in management perspective, they have strong relation with significance level of 0.01 and the causal Pearson

Malcom Criteria measurements	Leadership style strength	Strategic planning improvement	Customer and market focus management	Information and data analysis efficiency	Human resource focus	Process management	Business result capability
Leadership style strength	1	0.896**	0.960**	0.845**	0.855**	0.926**	0.929**
Strategic planning improvement		1	0.922**	0.820**	0.921**	0.861**	0.884**
Customer and market focus management			1	0.877**	0.882**	0.927**	0.958**
Information and data analysis efficiency				1	0.856**	0.843**	0.904**
Human resource focus					1	0.837**	0.881**
Process management						1	0.923**
Business result capability							1

***Correlation is significant at 0.01 level with Person (2-tailed) and list wise N = 100.*

Table 4.
Bivariate correlation in between Malcolm criteria measurement.

Correlation of most variable has a strong positive relation and their value is greater than 0.9 whereas the leadership vs. customer and market their value is 0.960 and it has highest value than the others which shows strong positive relationship.

6.3 Analysis of waste minimization tools measurement

Waste minimization tools measures are included six perspectives in this study. It identified each because waste issue is different from process angle and to see further correlation between each viewpoint. Waste minimization is the basic for any organization; the company does not adopt any particular standardized approach to larger improvement projects (**Table 5**).

When we see the above relationship, each variables in waste minimization tool perspective have strong relation with significance level at 0.01 and 0.05 whereas the causal Pearson Correlation of JIT vs. Six- sigma their value is −0.370 which show negative relationship and the highest Pearson correlation in waste minimization tool perspective is between lean vs. JIT its value is 0.920 which mean that JIT in NAS food is highest factor in Lean analysis.

6.4 Analysis of competitiveness measurement

Competitiveness measurements are included five perspectives that help evaluation and decision making within organizations that occupy in waste issue. It identified the correlation between quality, price, time, customer satisfaction and environmental views (**Table 6**).

When we see the above relationship of each variables in competitiveness measurement perspective they have strong relation with significance level of 0.01 and the causal Pearson Correlation of most variable has a strong positive relation and their value is greater than 0.9 whereas the time vs. customer satisfaction and

Factors	Lean	Just in time	TQM	Kaizen	Work Study	Six sigma
Lean	1	0.960**	0.042	0.858*	0.866**	0.335**
Just in time		1	0.0.34	0.858*	0.833**	0.37**
TQM			1	0.114**	0.244*	0.717**
Kaizen				1	0.868**	0.141
Work Study					1	019
Six sigma						1

*Correlation is significant at 0.05 level with Person (2-tailed) and list wise N = 100.
**Correlation is significant at 0.01 level.

Table 5.
Bivariate correlation in between waste minimization tools.

Do the following factors affect biscuit product	price affect competitiveness	quality affect the competitiveness	Time affect the competitiveness	Customer satisfaction affect the competitiveness	Environment affect the competitiveness
price affect competitiveness	1	0.898**	0.869**	0.803**	0.833**
quality affect the competitiveness		1	0.914**	0.895**	0.912**
Time affect the competitiveness			1	0.934**	0.940**
Customer satisfaction affect the competitiveness				1	0.913**
Environment affect the competitiveness					1

**Correlation is significant at 0.01 level with Person (2-tailed) and list wise N = 100.

Table 6.
Bivariate correlation in between competitiveness measurement.

environment their value is 0.934 & 0.940 respectively and it has highest value than the others which shows strong positive relationship it mean that time affect the competitiveness of biscuit product in NAS food is highest factor in customer satisfaction and environment analysis.

6.5 Analysis on waste level using the 7 lean wastes

There are certain techniques obtained from previous studies to analyses the seven lean wastes. Among them, the following stages were used.

- **Defining Stage**: The biscuit production process and determination of VA/NVA activities of NAS manufactures varies biscuits' production lines. The production process of line is run fulltime. According to many studies biscuit production and design revealed that the biscuit production process covers the stages of raw material preparation, mixing, forming or molding, baking by oven, cooling and packing [31, 32]. Each process has a certain design and layout in order to obtain quality, process capability and good capacity in order to meet the needs of consumers.

- **Measure Stage**: it a Waste Identification stage. During the field observation, the biscuit production process in NAS factory, there were several waste of resources identified i.e. non-standard process, fail on the ground,Crimean machine waste area, Rapper wastage, Packaging scrap product drops, error metal detector detection, broken, oval, overweight or small products, imperfect shape, non-standard water content, malfunction process, and engine breakdown. According to [33] Toyota identifies seven types of waste and they include 1. Overproduction, 2.waiting time, 3. Unnecessary transportation, 4. Excessive or erroneous processing, 5. Excessive inventory, 6. Unnecessary movement and 7. Defective product.to identifies the observed. The results of this identification were illustrated by a value stream mapping diagram, to determine the actual condition of the observed objects in several indicators, including value added and non-value added time. The value of Process Cycle Efficiency (PCE) was calculated to determine the value of Lean application level at NAS. Measuring stage is the process of measuring and identification of waste occurring at every stage of production process. The occurrence of each waste was measured and classified using the approach of 7-waste classification and finally calculated by Pareto analysis [34].

Analysis on the mapping process of the whole series of biscuit production is illustrated by some activities that are classified as non-value-added activities and some value-added activities (**Tables** 7 and **8**). Based on the time measure of the VA and NVA activities, the value of Process Cycle Efficiency (PCE) of 49.64% was obtained. The value of PCE is the result of division between Value Added Time and Total Cycle Time.

A company can be considered Lean if the ratio of value-to-waste ratio has reached a minimum of 30%; therefore, if the company is not lean and can be

Activity	Time(minute)
Preparation of flour material	5.0
Preparation of oil material	4.5
Preparation of packaging material	3.05
Weighing of other materials	11
Mixing process	20
Cutting & forming process	6
Baking process	4.5
Cooling process	5
Stacking process	7.8
Cream mixing process	10
Packing process	7.03
Cartoons process	4.8
Total	**78.88**

Table 7.
The value added process in the biscuit manufacturing for 3 month.

Activity	Time(Minute)
Lay time-dough	50.0
QC product check	30.0
Total	**80.0**

Table 8.
Non value added process in the biscuit manufacturing for 3 month.

categorized as a traditional company [8]. Because of the value of waste ratio is 36.7%.

- **Analysis Stage**: the definition and analysis of this stage is given as Determination of Critical to Quality (CTQ) and CPM value Critical to Quality. CTQ is a standardized or critical measure at every stage of production processes in order to produce quality products that meet the consumers' expectation in accordance with the capabilities of process technology available. Gazperzs [34] suggests that the characteristics of quality that will satisfy customers should first be identified. Here, the quality characteristics considered as critical should be classified and controlled. Each quality characteristic that has been classified should be determined to see whether it can be controlled through material, machines, work processes, and others control. CTQ standardization helps us to set up a maximum tolerance limit and a minimum tolerance limit. The values of USL and LSL are determining the process variation for each classified quality characteristic. They can also be used as signposts for product and process developments. According to the study by Hasan [35] stated that range of USL and LSL values is determined by the value of $\pm$ n sigma, and the Six-sigma approach (DMAIC method) is used as a reference in order to decrease waste or loss (**Table 9**).

As shown in **Figure 5**, the research found the values of six sigma calculation and enter the number defect observed is 5.25, enter the size of the sample are 78 and the defects per million (DPMO) of 67,308, and sigma of 3. This shows that the biscuit production of has a production capability with a failure of 67,308 every 1000,000 productions, or equivalent to 6.73% loss, and this indicates the production process still has a high failure rate. Also the research calculate DPOM, percentage of defect, percentage of yield, process sigma by process sigma calculator with inserting the number of defect observed and opportunities then automatically it calculate give result as we see above in the picture.

- **Improvement Stages:** this is the place where determination of FMEA is to be conducted and analyzed. A number of improvement steps were established at each stage of the existing processes from the preparation of raw materials, mixing, forming, baking, cooling, stacking and packing. Then, this stage tabulation was carried out on FMEA analysis. The FMEA method was used to determine the failure of the process and to analyze and improve the production quality [31, 36] (**Table 10**).

6.6 Qualitative result

Analysis on Baldrige model with six sigma methodology they state that "Baldrige provides the framework, Six Sigma the methodology."

Process stage	Critical to quality (CTQ)	Measurement	LSL	Target	USL
Preparation of raw materials	Process 1	Kg	6	6.5	7.1
	Process 2	Kg	10	10.5	11.3
	Process 3	Kg	12	12.5	13.2
Mixing	Process 4	Kg	31.5	32	32.7
	Process 5	Kg	199.5	200	200.8
	Process 6	Kg	89.5	90	90.7
Forming	Process 7	Gr	19.5	20	20.7
	Process 8	Gr	17.5	18	18.8
Oven	Process 9	Mm	49.5	50	50.6
	Process 10	Mm	44.5	45	45.7
	Process 11	%	2.5	3	3.8
	Process 12	PH	8.5	9	9.7
Cooling	Process 13	Wt.	39.5	40	40.6
	Process 14	Wt.	9.5	10	10.6
Stacking	Process 15	Gr	29.5	30	30.6
	Process 16	Mm	34.5	35	35.8
Packing	Process 17	Gr	139.5	140	140.7
	Process 18	Gr	29.5	30	30.8
Factors			PPM - Cpk Calculator		
Parts per million above USL			150		
Parts per million below LSL			−50		
Total PPM			100		
Zupper			5.12		
Zlower			6.25		
Z (Sigma)			5.22		
Cp			1.74		
Cpk			1.71		

Table 9.
CTQ of biscuit production process in one line in NAS food plc.

The experience of Motorola with Six Sigma helped the company to won the Baldrige award in 1988. According to Sumberg [37]; Parast [1] the Six Sigma quality laid the foundation for Motorola to be the first company to win the Baldrige award. Such a link between the Six Sigma methodology and the Baldrige model exists in practice.

The MBNQA framework has extended its application beyond businesses. According to the studies, it has specific guides for Education and Health Care organizations [38, 39]. MBNQA seven categories are Leadership, Strategic Planning, Customer and Market Focus, Measurement, Analysis and Knowledge Management; Human Resource Focus; Process Management; and Business Results [3]. Leadership shows how upper management chiefs the organization and organization community. The strategic planning is also the organization establishment of plans to

Figure 5.
Control chart for UCL and LCL.

implement strategic directions. Customer and Market Focus is the organization that builds and maintains strong, lasting relationships within customers. Measurement, Analysis and Knowledge Management are also the organizations use of data to support key processes and manage performance while human resource focus is outlining of the importance of human resources. Process Management uses powerful tools like Continuous improvement program, Zero Defect, and Re-engineering. Continuous improvement recognizes that, even when no errors occur, there are opportunities to improve the design of the process or product. All the time, the competitors are seeking to gain an advantage by making their products better. If the companies do not seek to improve, it will get left behind. Company should expect to receive no complaints from customers. This goes beyond the idea of keeping complaints to a minimum. It indicates that the company should adopt a new approach, perhaps checking that each customer is satisfied with his purchase [40]. Finally, business results is an indication of the organization improve in terms of customer satisfaction, finances, human resources, supplier and partner performance, operations, governance and social responsibility and how the organization compares to its competitors.

6.7 Proposed continuous quality improvement model

The integration lean, six sigma and Baldrige model understand separately and define similarity and dissimilarity in each alignment some criteria are considered based on the literature review. These include: focus on work condition in addition to process, easily understandable and having proper metrics; addressing the possible root cause of continuous quality improvement problems; being continuous improvement tool; having clearly defined improvement goal; flexibility to apply from operation to firm level and involving all, including top management to low.

6.8 Feature of the proposed model

6.8.1 Applying six sigma in the lean philosophy

It is true that there is established link between Six Sigma methodology and the lean manufacturing with in the previous studies. The proposed integrative six sigma and lean manufacturing is based on the following basic pillar principles. These principles are:

- The Six-Sigma methodology is linked with the lean. It becomes part of the Lean six sigma (LSS) for achieving waste and defect reduction techniques.

DOI: http://dx.doi.org/10.5772/intechopen.95279

Processes Biscuit making	Steps/ Inputs	Failure mode	Failure Effects	SEV	Causes of failure mode	OCC	Current control	DET	RPN
Material preparation	Material weighing	Wrong ingredient	Inconsistent quality	10	Substandard material supplied by supplier	5	Undetectable	10	500
		No electricity power	Damaged dough	8	Main Distributor Panel Trip	6	No monitoring	10	480
		Blocking wheat at hopper	Line stop	7	Damaged Filter	6	No monitoring	10	420
Mixing Process	Mixing	Different usage of water between shifts	Unstable dough and waste	6	Differences in methods by operators	10	Controlling only after mixing is finished	6	360
		Frequently error censor	Down time and safety issue	6	Unstable socket pad	10	Alarm system of mixer is on	5	300
Forming Process	Forming	Product Tailing	Product jam	10	Uneven surface of blades	10	Checklist of forming	6	600
		Product Tailing	Product Jam	10	Loose molder Teflon	10	Visual control	5	500
Oven Process	Baking Process	Irregular thickness of biscuit	Unbalanced thickness of biscuits	10	Uneven weight among rows	10	Check list of intense monitoring	5	500
		Product Jam	One side of Wire mesh is loose	10	One side of Wire mesh is loose	10	Visual control	7	700
Cooling process	Cooler	Product is Excess Calm	More rejected products	10	Scrap was taken accidentally	10	Only Visual	5	500
Stacking process	Sandwich	Thin or thick biscuits	packing machine often starts or stops	10	Cream texture is different since the icing weight is different	10	No control/ monitor	10	1000
	Mixing cream	Unbalanced Cream	Waste/ thin or thick	10	Inaccurate balance or weighing scale	10	Inappropriate display	8	800
Packing Process	Product packaging	Broken Products	Dead machine, quality potential	10	No sorting tool of sandwich since taken accidentally to packing	10	Visual	10	1000
		Poor packaging material	Inconsistent quality	10	Substandard material supplied by supplier	6	No control	10	600

Table 10.
Improvement process 5) control stage: Plan to continue measuring the success of the updated process is usually created and any documentation, process or training material is updated.

- The lean self-assessment includes the assessment of each seven waste in the six-sigma (DMAIC).

- The proposed model for the integrative lean Six-Sigma is capable of addressing the core values of the lean. Areas such as overproduction, waiting time, unnecessary transportation, excessive or erroneous processing, excessive inventory, unnecessary movement and defect can be addressed by Six Sigma methodology.

6.8.2 Applying Six Sigma in the Baldrige Model

It has been established a link between six sigma methodology and the Baldrige model in similar fashion as six-sigma with leans. The proposed integrative six sigma and Baldrige model given in **Figure 6** is based on the following principles.

- The Six-Sigma methodology is linked with the Baldrige model. It becomes part of the Baldrige model which is considered as a single unit for achieving performance excellence. This performance analysis requirement is set by the top management. After the establishment of Baldrige model goals, Six Sigma methodologies are used to increase the processes and meet quality purposes.

- Six-Sigma projects can be applied to all types of the projects, processes, and products. Based on this application, the selection, administration, and control mechanisms were directed by the top management used in this proposed model.

- The proposed model for the integrative six-sigma-Baldrige Quality in **Figure 6** is capable of addressing the core values of the Baldrige model. Areas like

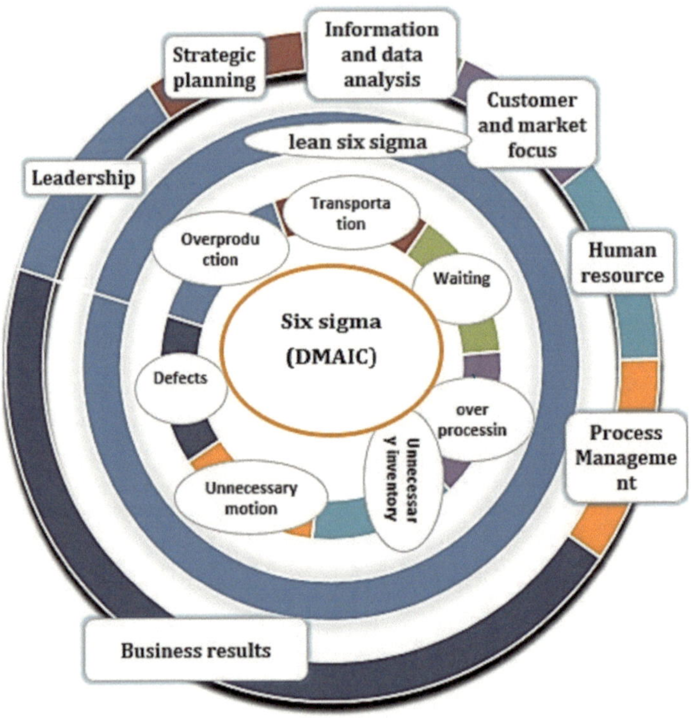

Figure 6.
Integrated lean-six sigma model.

leadership competencies, strategic development and deployment, and human resource management can be addressed by six-sigma methodology which is a powerful tool.

6.8.3 Applying six sigma in the Baldrige model and lean

We established a link between six sigma methodology and the Baldrige model also six -sigma and lean. The proposed integrative lean Six Sigma and Baldrige model is based on the following principles:

When lean, six sigma integrated with Baldrige model. In fact, it becomes part of Baldrige model for achieving excellence. Such a performance requirement is set by the top management. Afterward aligned with the requirement of the Baldrige model, lean and six sigma methodologies become used for improve the process and meet quality objective.

The Baldrige self-assessment includes both the assessment of each seven categories in the Baldrige model as well as the efficiency and effectiveness of the six sigma projects. Such an approach toward six sigma and lean projects ensures that the company is gaining benefit from implementing the lean- six sigma methodologies.

The proposed models for the integration of lean, six sigma with Baldrige is capable of addressing the core value of the Baldrige model and identify the seven waste of lean. Area such as leadership competencies, strategic development and human resource management can be address by six sigma methodology whereas for the integrative lean Six Sigma identifies such area overproduction, waiting time, unnecessary transportation, excessive or erroneous processing, excessive inventory, unnecessary movement and defect can be addressed by Six Sigma (DMAIC) methodology.

7. Conclusion

The conclusion of this research is that the waste minimization at NAS food Plc. as production process applying the lean process. In such cases, the standard DMAIC cycle may provide structure and ensure that each step is improve thoroughly, thereby helping the success of the project. Lean can contribute to these projects by staking out the direction; that is, indicating where to start, for example through the use of value stream mapping where the process is reviewed in order to find waste. Analysis on the mapping process of the whole series of biscuit production is illustrated by some activities that are classified as non-value-added activities and some value-added activities.

In this paper both qualitative and quantitative analysis are applied and the result of qualitative by using (DMAIC) based on the time measure of the VA and NVA activities, the value of Process Cycle Efficiency (PCE) of 49.64% was obtained. And the research found the value of waste ratio is 36.7%, this show non lean in the company and the values of six sigma calculation and insert the number defect observed is 5.25, the size of the sample are 78 and the result defects per million (DPMO) is 67,308 and sigma of 3. This shows that the biscuit production of has a production capability with a failure of 67,308 every 1000,000 productions, or equivalent to 6.73% loss, and this indicates the production process still has a high failure rate. And improve by using FMEA analysis was carried out each stage of the existing process used to determine the failure of the process and to analyze and improve the production quality and the result of highest RPN on Stacking and packaging process so they should take action for waste minimization and continuous quality improvement. In addition, Organizations try to implement the Baldrige model as a means for achieving excellence. The seven categories within the Baldrige model is integrated and related with the purpose of addressing quality challenges so that companies can be competitive in the dynamic business environment.

In quantitative analysis used SPSS software and the Correlation and regression analysis is used to quantify the association between two continuous variables (between an independent and a dependent variable or between two independent variables). The result of this shown relationship of each variables in waste measurement perspective they have strong relation with significance level of 0.01 and the causal Pearson Correlation of most variable has a strong positive relation and their value is greater than 0.8, likewise The good regression results of equation is selected in each section which is explained R2 value about 0.9 or 90% with statistically significant ($P < 0.01$) level.

However, In this paper, relying on a review of the national and international literature, 25 key indicators was selected based on the correlation having strong 'r' value under four improvement perspectives were identified to assess measurement system of the organization. The Lean Six Sigma & Baldrige application studied here does not point toward one well-defined Lean Six Sigma approach; the company does not adopt any particular standardized approach to larger improvement projects. Instead, the company supports the integration at this level by ensuring that their improvement specialists are widely trained in Lean, Six Sigma and Baldrige model.

8. Recommendation

Further research is required for biscuit manufacturing in terms of waste minimization at each process stage using this DMAIC methodology. This study needs to be continued in terms of FMEA usage along with the development selection model for waste minimization and improving the production process. Instead, the company supports the integration at this level by ensuring that their improvement specialists are widely trained in Lean, Six Sigma and Baldrige model.

Similar research also should be conducted in order to provide added value in the fields of food or agriculture with the application of lean six-sigma and Baldrige model.

Conflict of interest

The authors declare no conflict of interest.

Author details

Kassu Jilcha Sileyew* and Selamawit Gebreyohanis
School of Mechanical and Industrial Engineering, Addis Ababa Institute of Technology, Addis Ababa University, Addis Ababa, Ethiopia

*Address all correspondence to: jkassu@gmail.com

References

[1] Mahour M.Parast, E. C. (2006). A Relationship between Six Sigma And Malcom Baldrige Quality Award.

[2] Coskun Dalgiç, H. V. (2011). Improvement of Food Safety and Quality by Statistical Process Control (SPC) in Food Processing Systems: A Case Study of Traditional Sucuk (Sausage) Processing. InTech, Available from: http://www.intechopen.com/books.

[3] Png, C. L. (2015). A Review of Lean Six Sigma and Malcolm Baldrige. Volume 6.

[4] Mousa, A. (2013). Lean, six sigma and lean six sigma Overview. International Journal of Scientific & Engineering Research, volume 4(5).

[5] Liang Y, et al. (2014). Engineered pent functional minicellulosome for simultaneous scarification and ethanol fermentation in Saccharomyces cerevisiae. Appl Environ Microbiol 80 (21):6677-6684

[6] Alie Wube Dametew, D. K. (2017). The Roles of TQM and JIT for Basic Metal Industries Global Competitiveness. Industrial Engineering & Management, 6(2).

[7] Workneh, H. (2010). Analysis of Technical Efficiency of the Ethiopian Agro-Processing Industry: The Case of Biscuit and Pasta Processing Firms. AAU School of Economics.

[8] Ibrahim Alhuraish, Christian Robledo, Abdessamad Kobi (2017), A comparative exploration of lean manufacturing and six sigma in terms of their critical success factors, Journal of Cleaner Production, Volume 164, 15, Pages 325-337

[9] Yasin, a. (2014). Application of lean six sigma for process improvement: the case of Ethiopian paper and pulp s.c. Addis Abab: AAiT.

[10] Imam, H. (2010). Recycling in a supply chain context - A case study regarding sorting and collection. Melardeen University of Seweden.

[11] Pap, N., Pongrácz, E., & Myllykoski, L. (2014). Waste Minimization and Utilization in the Food Industry. Jatindra Kumar Sahu (ed.).

[12] Dave, N. (2002). How to compare six-sigma, lean and the theory of constraints. Volume 6.

[13] Eshetu, B. (2017). Integrated Model for Continuous Productivity Improvement in Footwear Industry: (A Case of Anbessa Shoe S.C.). AAiT.

[14] Kumar, R., & Kumar, V. (2012). Lean manufacturing: elements and its benefits for manufacturing industry. Trends and Advances in Mechanical Engineering,YMCA University of Science & Technology, Faridabad, Haryana.

[15] Thomas Pyzdek (2003),The Six Sigma Handbook, A Complete Guide for Green Belts, Black Belts, and Managers at All Levels, McGraw-Hill Companies, T.H.E.

[16] Womack, J.P., Jones, D.T. and Roos, D. (1990), “The machine that changed the world: the story of lean production”, How Japan’s Secret Weapon in the Global Auto Wars Will Revolutionize Western Industry, Rawson Associates, New York, NY

[17] Ron Snee (2010), Lean Six Sigma – getting better all the time, International Journal of Lean Six Sigma 1(1):9-29 DOI: 10.1108/20401461011033130

[18] Eckes, G. (2001),The Six Sigma Revolution, Wiley, New York, NY

[19] Andersson, R., Eriksson, H., and Torstensson, H. (2006) Similarities and differences between TQM, Six Sigma and Lean, The TQM Magazine, 18(3), 282-296.

[20] Vokurka, R. Stading, G. & Brazeal, J. (2000). A comparative analysis of national and regional quality programs. Quality Progress, pg 41-49

[21] Singh, C.D., Singh, R. and Singh, S. (2013), "Application of lean and JIT principles in supply chain management", Journal of Regional and Socio-Economic

[22] Horne, J. R. (2009). The effect on corporate performance of firms that won the. H. Wayne Huizenga School of Business and Entrepreneurship Nova Southeastern University.

[23] Nave, D. (2002). How to Compare Six Sigma, Lean and the Theory of Constraints. USA: American Society for Quality.

[24] Ahmed, M. (2013). Lean, six sigma and lean six sigma. International Journal of Scientific & Engineering Research, 6 (2), 1137-1153.

[25] Hans, S. B. (2013). Applying Lean, Six Sigma, BPM, and SOA to Drive Business Results (Vol. 6). IBM.

[26] Young, B. (2001). Waste Minimisation An Environmental Good Practice Guide for Industry. The Environment Agency.

[27] Talib, F. R. (2011). Total quality management and service quality: an exploratory study of management practices and barriers in service industries. International Journal of Services and Operations Management, 10(1), 94-118.

[28] Jangwoo Lee, K. L. (2015). Supplier Partnership Strategy And Global Competitiveness:A Case Of Samsung Electronics. Eurasian Journal of Business and Management, 3(4), 1-12.

[29] Walder, J. (2007). Integrated Lean thinking & ergonomics utilizing material handling assist device solutions for aprodction work place.

[30] Sunilkumar N Chaudhari1, A. J. (2015). JIT Implements in manufacturing industry – A Review. International Journal of Engineering Research and General Science, 3(4).

[31] Ahmed M, s. a. (2017). PT kraft ultrajaya indonesia. indonesian journal of business and enterpreneur, 77-89.

[32] Link, A. N. (2001). Economic Evaluation of the Baldrige National Quality Program. University of North Carolina at Greensboro.

[33] Womack, J. a. (2003). The Machine that Changed the World. New York.

[34] Addae-Korankye. (2013). Total quality management (tqm): a source of competitive advantage. A comparative study of manufacturing and service firms in ghana. International Journal of Asian Social Science, 3(6), 1293-1305.

[35] Milovanoi, M. (2011). Just in time concept as a mean for achieving competitive advantage in the virtual economy. In b. Katalinic (Ed.). 22. DAAAM International, Vienna, Austria, EU.

[36] Md. Mazedul Islam, A. M. (2013). Application of Lean Manufacturing to Higher Productivity in the Apparel Industry in Bangladesh. International Journal of Scientific & Engineering Research, volume 10.

[37] Praveen, S. (2012). Reduction of working in process inventory & production lead time in bearing industry using value stream mapping. volume 9.

[38] Talib, F. (2013). An Overview of Total Quality Management:

Understanding the Fundamentals in Service Organization. International Journal of Advanced Quality Management, 1(1), 1-20.

[39] Kassuand Kitaw. (2016). A Literature Review On Global Occupational Safety And Health Practice & Accidents Severity. International Journal for Quality Research, PP 279–310.

[40] Dr. H. Nagaprasad, 1. B. (2008). Making World Class Product Through Quality Process Management.

Chapter 11

Analysis, an Anathema: Is That a Fervent Diatribe of Lean?

Sajit Jacob and Krishnamurthy Kothandaraman

Abstract

Should there be an understanding that rigor in analysis must be out-of-bounds for Lean initiatives? Will this rigor not facilitate a benchmarking of Lean initiatives? Why not a Lean initiative cause-consequence assessment not performed for building future fault tolerance? The effectiveness of a company's strategy is critical to its success or failure. Lean strategy seems to be claimed as a widely recognized factor for business success and competitive advantage. However, empirical evidences do not promote the idea that Lean has delivered results every time. Study results indicate that success or failure of lean initiatives strongly depends on how companies approach it and on whether company has created their own curated philosophy towards Lean. Then, success is not dependent alone on a strategy, but on how daily operations are aligned to strategy. This chapter aims to address the above questions and a greater number of questions that we experience on a day-to-day basis with regard to Lean applications in the real world. Chapter Learning Objectives: Understanding Lean, Lean failure modes, and Lean initiative precautions.

Keywords: Lean, failures, assessment, arguments, 6Cs

1. Introduction

Why Lean initiatives are not analyzed independently and collectively to understand the failure modes that resulted in many failures firms conceded in the past? Why Lean is more 'appealing' to the corporate leadership when pros and cons of the methodology and its nuances are not well studied? Has Lean ended up as one of the many continuous improvement initiatives many organizations have undertaken as it does not demand long rigorous trainings, no expectation of quantitative acumen, no requirement of good historical data, decision on the effectiveness and efficiency outcome is completely within the ambit of the enterprise giving the ultimate flexibility?. Did absence of a structured methodology, benchmark and a third-party assessment has given the maximum convenience and performance priority low? By giving fancy terminologies such as transformation, high velocity development, out of the box idea generation, have we lost the direction and purpose?

Should there be an understanding that rigor in analysis must be out-of-bounds in lean philosophy? Is that absence makes Lean an affable, acceptable, and appealing slogan to a larger section during their attempts to cross the barriers in eagerness to reach the holy grail of excellence by quicker means. In total irreverence, if lean has to fail what more could be a bigger reason, when this strategy is recognized for the questionable characteristics in pursuit of agility, such as superficial management, reluctance to examine sustenance of accumulated benefits, and avoidance of retrospective studies. So is Analytics, an inevitable villain, in the drive to excellence? Process mining and data analytics are integral to business excellence journey riding on and capitalizing the benefits of Lean, augmenting the methodology without missing its innate flavor. The new normal induced today's global economy, characterizes demand specificity, spend thrift consumers, substitute products, aggressive pricing, etc. has created a breed of customers who are demanding much more than ever. The innovate and compete has become imperative and inevitable norm of the day. Improve to sustain and survive, but not at the cost of bleeding reserves, rather by optimization and conserving. This can only be done by minimizing the manufacturing cost of products by increasing the productivity and reducing wastage during production. Therefore, the industrial organizations realized the need for efficient and effective use of resources in a way that justifies production economics [1]. Thus, these organizations tried adopting and adapting several strategies to confront this challenge, including the lean manufacturing strategy [2].

The concept of lean manufacturing originated in Japan with an intent to conserve funds by eliminating wastes by identifying sources of waste and then using tools to eliminate them. It is now widely publicized that organizations that practices lean manufacturing methods produces world class products that have significant cost and quality advantages over those who still practices traditional mass production. But, if we have to claim, Lean has its origin in Toyota Production system, then we also need to agree that the system at Toyota is integral to Toyota way of life and if another company has to replicate the same success they need to develop their own unique values, principles and priorities of life. The Toyota philosophy has evolved over a period of time over a value system that thrives on safety, security and motivation of their work force. Thus, Lean has to be a way of life that is unique to an implementation and cannot be a medication only at a time of illness rather it is a vaccination schedule for a life.

It is concluded from the available literature that the lean techniques are theoretically applicable in all industries and has proven their success in practice specifically in large organizations. It is the management style that sets the tone for employee attitude that determine the maturity of lean operations within a company and they set the culture of the lean organization [3]. The lean environment that takes undue advantage of the flexibility lean offers and the fear psychosis instilled by leadership to find a waste as a mandatory dictum together works counter-productive to leaning operations.

Despite its long existence, Lean has been moving with the tide set-up by socio-economic and political winds that prevailed at those points in time in the enterprise. Post world war2, the demands varied through the years as countries slowly regained their economic stability. During this time, the challenges determined, for what Lean must stand for. At times, it was shortage of skilled men and raw material then demand from Lean was optimum resource utilization, scarcity of storage houses forced to have zero inventory as a target. As economy got its boost, commerce benefited and demand certainly began to rise, then managing supply versus demand became a factor that created market advantage, hence Lean focused on quicker delivery with minimal steps to produce. Labor unrest, famines, pandemics and cost highlighted the need for Lean to focus on human resource management. Once commerce flourished and alternate products flooded, Lean turned attention to meeting productivity targets at reduced cost. Thus, during the tumultuous post war period, Lean revised, and improved its definitions, multiple times.

Overall, the management commitment, financial sponsorship, competency development, and culture; probably are majorly impacting lean operations. The rest of the chapter proceeds with a literature review that identifies the different perspectives evolved in the prior papers. Then, research methodology explains the method adopted by the authors to complete this study. Then comes, arguments and discussion that outlines the various failure themes and then, there is an outcome recommendation that reflects the possible procedural precautions that may control the recurrence of potential failure modes, and finally chapter culminates in the concluding thoughts of authors.

2. Literature review

Lean is a combination of principles, practices, tools, and techniques with an aim to improve safety quality, cost, delivery, productivity and improvement by eliminating non-value adding steps. Further, lean is a continuous improvement initiative with an intent to implement business processes that with minimal waste and reduced lead times [4]. While elimination of wastes and direct implication on value perceived by the customer are heavily promoted, most of the success stories in Lean originates from Japan, while many of the failure stories finds their way from rest of the world [5].

With the decision of an organization to initiate the lean process comes the challenge of bringing about the change in the thought process of employees and work culture [6]. This is because lean is a way of working towards the elimination of waste across organization, thus a transition of behavior and methodology that may be deeply rooted within an organization is required. When an organization chooses to go lean, it hunts for waste across system, thus earns the distinction of being a socio-techno intervention.

A trust-based work culture is a precondition for lean intervention as leaning raises the anxiety of job loss in employees. The improvement process must be recognized as benefiting both the company and the employees. The ultimate responsibility for the outcome rests with the management. Thus, studies highlight that the major roadblock in successful implementation of lean manufacturing that lead to improvement in production effectiveness is the hesitation of management to empower employees. Ahuja and Khamba studies also share the same viewpoint that the rigid bureaucratic structures of the organizations are impeding empowerment of the employees. Lack of employee involvement in the overall implementation can lead either to their failure or partial implementation of these systems. Based on the above discussion, the following arguments has been formulated [7].

Unsuccessful implementation of lean manufacturing techniques is caused by employees' reluctance, lack of training and ethical education, and lack of follow up by the officials in the organization [8]. The purpose behind training and retraining of employees is to develop multi skills that could help them work more diligently, enthusiastically, independently and responsibly [9].

It is important to identify the causes for failure and understand their reasons and implications to assure a minimum probability of success in subsequent ventures with Lean [10]. The barriers to lean implementation can be grouped into the following ten broad areas by characteristics: organizational culture, knowledge, management, conflict, resources, technology, finance, employees, customers, and past experience [11]. Most of the occasions, a failure in Lean implementation is being attributed to lack of knowledge in lean, impatience and ignorance of the benefits being reaped by a cross section of companies leading jeopardizing the implementation at different stages. Isolated implementations as well cannot bring sustainable developments in performance even if the method is Lean [12]. Using the know-how on Lean in appropriate circumstances in applicable industries will reap better rewards than a mass application [13].

3. Research methodology

Case research is particularly valuable when the intention is to examine phenomena in their natural setting. In addition, according to Rivera and Chen studies [14], case studies are appropriate when the research seeks to address "how" and "why" questions. The type of case research employed in this study is a retrospective case study of ten companies. This research perspective enables a thorough, in-depth analysis of the various aspects involved in the adoption of Lean strategy, by examining retrospective views of an unsuccessful attempt to implement Lean strategy. A major benefit of a retrospective approach is the reliability of the case's selection, since the sustainability, of strategy implementation can only be evaluated in retrospect [15]. Inaccuracies in artifacts, interpretation and perceptions, priorities, and objectives are influencers in this method.

Author was either a listener or observer in the process of understanding lean thinking perspective and implementation styles across various organizations. While conducting multiple case studies, benchmarking of cases with theory and inter cases comparisons were conducted to understand the environmental differences. According to the multiple case study method evaluated, the sufficient number of the cases required for this study is envisaged as 10 Lean failures. To assess the companies approaches and results of lean implementation; data collection step was performed. The types of data collected were selections, narrations and visual experience.

Assessment of the companies consisted of three main parts: assessment of lean adoption steps, perceived performance of proposed processes and actual results of those processes in reality; and degree of Lean implementation and institutionalization.

Lean adoption was evaluated based on the status of defined protocol that also includes work environment, management, performance analytics, competency, work force morale, risk and continuity, and change handling aspects as well, as its effectiveness and efficiency indicators of maturity, and finally practice evidence. Perceived performance of process is evaluated by the estimation methodology defined, and its application evidences. Actual results are the observed values and its practice evidence in the form of data collection formats and associated practices. Degree of implementation and institutionalization maturity was assessed based on the simplified model [16]. According to this simplified model, nine criteria of lean implementation maturity are assessed: elimination of waste, continuous improvement, zero defects, just-in-time deliveries, pull-of-raw materials, multifunctional teams, decentralization, integration of functions, and vertical information systems. Each criterion has determinants. Determinants describe the results of implementation of corresponding criterion. Determinants are assessed with explicit rules of coding such as, 2 for well implemented, 1 for fairly implemented and 0 for poorly implemented. Those grades are brought in by author. Such grades are chosen from simplicity point of view and only with the aim of classifying content based on the degree of existence or a peculiarity of a particular characteristic in data. Assessment is made by comparing the initial state before lean initiative started and the state of each area by the time of assessment. Similarly, specific to the assessment of lean implementation results; determinants were summarized, and qualitatively compared and quantitative translation and summarization was avoided to prevent this paper from drifting towards a biased conclusion, rather messages must be presented to further the possibilities of a balanced quantitative research.

Collected and classified qualitative content in the form of text, narrative and visuals were analyzed by using content analysis method. Content analysis method could incorporate the various kinds of analysis where communication content is categorized and further classified and is a systematic, replicable technique for compressing many words of text into fewer content categories based on explicit rules of coding. Data analysis in current paper used the coding approach after following some preliminary examination of the data: material is reviewed and a set of features in the form of checklist is created, further applied for coding. As such, validation of the inferences made on the basis of data from one analytic approach demands the use of multiple sources of information. Meaning, the researcher should try to have some sort of validation study built into the design, for example in the form of triangulation, which is often used in qualitative research. By triangulation the credibility of the findings could be achieved by incorporating multiple sources of data. In current research three main types of data were used. Based on the content analysis method, the data was naturally categorized based on criteria from Karlsson and Ahlstrom (1996) model. Next, data were analyzed and concentrate of needed information were brought out based on data type – text (company documents), narrative (questionnaire and interviews) and visual (photos, video and field notes). Further, summarizations were given to each determinant based on data available.

The summarized qualitative information is compartmentalized into twelve themes that forms the twelve Lean arguments, around which an argumentative approach was facilitated. In this approach, the well promoted stand on Lean is exposed based on the content analysis inferences.

4. Arguments and discussion

Argument 1: A method of success-by-design.

Lean promotes the advantages of developing quality, productivity, reduced inventory, flat structures, teaming and flexibility [17]. Main principles of Lean: specify value from the standpoint of the customer, identify all the steps in the value stream and relentlessly work towards eliminating steps that do not create value, make the steps flow smoothly towards the customer, let customers pull value from the next upstream activity and begin the process again until a state of perfection is reached. Lean in the absence of a repeatable and reproducible methodology in reality is a method of success-by-chance with tools and techniques. Lean does not believe in investing in elaborate designs of experiments or benchmarking due to the absence of standard methodology and measurements.

Lean promotes an organizational transformational context; therefore size, age, complexity, infrastructure and competency create the essential environment that decides the strategy of the endeavor. Then the change management determines the success of the initiative as it is bestowed with the charge of change over, risk control, sustenance, disaster recovery and business continuity while transforming the situation [18]. Thus, observations favor the counter argument, that Lean is a method that provides success-by-chance.

Argument 2: A method that favors industrial relations.

Teaming for the enterprise is different from democratic industrial relations. Lean teams are with limited objective and scope of operation. The teaming process in Lean revolves around, a process or a machine or an area of the production floor. The competency required to be part of the team is limited to lean tools. Thus, Lean rhetoric is countered in this argumentation with a selected scope and interpretation.

Argument 3: A method aligned to economics.

In Neo-liberalized and capitalist economy, continuous improvement through waste elimination to eventually generate value is not a choice. Classic example is the growth of IT industry in the developing countries on the cusp of maintenance, enhancement and support projects of many softwares that were developed earlier. Thus, at times, waste itself creates an industry, then that waste becomes value adding waste. Any resource that does not generate value to customer is eligible to be called a waste [19] is not a sustainable argument as for one service provider, an effort that could be termed as waste may eventually create a new business line, more employment and an expansion in the purchasing power of society. Thus, Lean need not find a complete alignment with social economics.

Surveys claim, lion-share of lean initiatives fail to achieve systematic productivity improvement and even suspects the benefits claimed in many of such initiatives [20]. Popular as a management system and has become a style symbol of operations, as part of leaning operations may save for the organization, but reduction in manpower and remuneration restructuring to the lower side associated with such initiatives impacts the purchasing power of the society. This indirectly points to disparate definitions and perceptions in interpreting the claims.

Free-market economies propounding laissez-faire economics that allows a free hand to the business to devise norms to conquer market with ruthless competitiveness and reduced subsidies for the sake of survival was forced to shed virulent trade union intransigence leading to a social and political transformation. It is essential to investigate the role of Lean methodology in promoting unemployment in many industrial belts. Unchecked capitalism and globalization have taken away pluralism in management and turned focus to discover value for the customer as an outcome of value chain. This is possible only by increasing productivity at lower costs, so internally, the focus is still not on customer but on production systems. If value and loyalty are perceived in faith, the politics in displayed improvements comes obvious.

Argument 4: A method that sustains benefit.

The repeatability and reproducibility of the results associated with Lean initiatives are invariably absent in the survey. It also could be due to absence of a structured methodology, accepted glossary of terminologies and availability of verifiable results. Thus, absence of a standardized scientific method creates a void while treating lean as a system to enhance business practice rather than industry sectoral differences.

The experiences from Toyota Production System of 1970s formed the basis of western attempts to build a corpus of derived knowledge out of it to create a book of knowledge on Lean [21]. The whole objective of Womack et.al [22, 23] in 1990s was to create a theoretical basis applicable irrespective of industry to gain advantages that are agreeable irrespective of the verifier. The hypocritical bias and judgmental attitude took a backseat as soon as Lean terminology emerged out of International Motor Vehicle Program. Eventually, five principles that laid the foundation of Lean emerged and they were concept of value, value stream mapping, perfecting the ideology of pull the flow. Then it was a time for a plethora of techniques to evolve along with contexts for their application.

Lean is industry independent, while its roots may be in production, and it is prevalent in services as well. Lack of empirical rigor is an established issue in Lean and even research papers are analytically inadequate as most authors prefers to discuss around an event, experiences or an implementation and rarely it gets into compare and contrast analysis of a cluster of failed case or case surveys to reveal the differences, factors involved, extent of success that can be attributable to plans, implementation models, and actions. Even qualitative analysis when it is contextual it generates arguments, possibilities, hypothesis which could eventually be tested to identify the effectiveness of the Lean effort, which rarely happens [24]. While lean remains as a collection of techniques, their relative importance and effectiveness and efficiency of performance are yet to be quantitatively evaluated to structure it into a framework or lifecycle to give universal acceptability, thus its ability to provide a sustained benefit cannot be proved beyond doubt.

Argument 5: A method that sustains business.

Radnor and Johnston studies [25] is of the view that lean transformation mainly motivated by the cost reduction associated with it rather than the customer value it can bring. But at times waste also sustains business as is the case with certain failures create a demand to do the right for the customer which becomes the subsequent version that would sustain the life cycle of a product in the market and the antithesis is it is the consequence of a poor service design. So then what is a leaning operation and how to perform a lean service. Unless there is a clarity, a framework, and process and tools to support it and metrics to measure, the claims will become redundant. Else, adopting and adapting techniques that showed result elsewhere when planted in a different environment and in different context, need not realize the same outcome. In other words, techniques that provided benefits in a industry under specific conditions need not reward the same way, and it may even import unwanted effects [26].

How much pull effect can reverse low demand in a production environment? Pull effect is on the basis of creating a product that generates a desire in a customer which translates to a product demand. Demand is not merely based on features, but also determined by affordability, necessity, substitute and uniqueness that together generates the value the product generates in the market. Lean is not a product innovation methodology, rather it always remains as method to optimize cost. Unless cost is balanced with features and ego satisfiers, the product will meet with the history of Nano car in India. Hence, lean is never a guarantee for a product to be a success in the market.

Argument 6: A methods that is universally applicable.

Lean, because of its origin in manufacturing may have all the wastes defined with a manufacturing perspective, then a direct correspondence in services is difficult to find. For example, a defect in a tangible product and a defect in service is totally disparate. Similarly, the context determines the definition of defect, for example, an excess inventory in manufacturing stock foretells possible wastage, but at the same time a shortage in inventory need not be ideal too as it may create opportunity cost in services and may be appreciated as JIT in manufacturing [27]. So universality of the concept is at the goal level that is, generate value by reducing waste rather than debating at applicability of the definition of the type of waste in different industries and sectors, moreover, in service value is co-created while in manufacturing producer creates and recipient perceives. This becomes clear when it is translated to monitory values where post-service revenue is based on conditions created by customer and hence lack definiteness, but a tangible product revenue is definite and earned as soon as it is sold. This brings the curtain down over multi-sectoral comparability. Even in strategy, a pull method is a possibility in manufacturing but the same cannot be expected in services as arrival rate and pattern in services is a matter of probability.

In all safety critical industries such as health, aeronautics, nuclear energy, oil and natural gas etc., active Lean employment on that system is not advisable. The situation is identical in every production space where there is a statutory and regulatory compliance involved. But any other Lean efforts performed in other departments, if it reduces environmental impact score, then that will be considered as a financial incentive generated out of resource conservation and pollution reduction. Thus, there are scenarios where Lean is not likely to be the drug-of-choice.

Argument 7: A method neutral to production peculiarities.

While Six Sigma strives for near perfection, Lean tries to accelerate the velocity of the end-to-end process by reducing the Lead Time. But consistency in achieving near perfection at high productivity rates is not mere LSS achievement. Lean promotes the idea of high rate of production in small batches is under the assumption of Overall Equipment Effectiveness and maintaining constancy of quality is never a guarantee.

In those industries where bulk production is not possible, Lean faces limitations. In case of multi service counters, made-to-order large engineering goods, where demands are unique for every customer. Under such conditions, attempting Lean at higher levels where commonality can be perceived is tried as an option, but farther in the value chain when optimization strategies are attempted, the real value perceived at the recipient end is limited.

Then the methodology claims to have the ability to maximize shareholder value by consistently improving Quality, Cost and Customer Satisfaction. Absence of credible quantitative data to evaluate and benchmark performance of lean projects is unavailable is a major constraint. If it has to be proved that it is the methodology that makes difference, it is equally significant to nullify the effects of factors such as industry, culture, competency, production volume, process, technology, material involvement etc. The performance that are pertinent are, but not limited to the following areas such as lead time, waste management, labor productivity, and economic value add synergistically. Is yet to be ascertained consistency in performance, causality for variability etc. Thus nature of product and volume could be critical determinants while deciding on Lean as the method for process improvement.

Argument 8: A method in itself is a strategy and enhances creativity.

While subjectivity, scale and understanding affects the measurement system error; an authentic survey analytics to agree on source or type or excellence-based clusters formed by practices, techniques, and tools is still not available. Principal Component Analysis with Varimax rotation will generate principal component with high factor loadings to identify a theme associated with it that significantly influences the success of a Lean initiative. Does Lean promote innovation? Role of lean in creativity through product designs and functionality is a questionable character, hence its strategic role is of limited extent. Success of a operational strategy must be linked to performance parameters such as quality, cost, rate of flow, safety, and innovation. If Lean has to qualify itself as a strategy, then it needs to perform on all such parameters to stay in isolation as a independent methodology [28].

Argument 9: A method with many techniques as strength.

In lean, plenty is the problem when it comes to techniques. Then, appropriateness in choice and correctness in application determines fate of the problem. Deciding on which of the many lean tools to apply, where and when and how it is applied create the inevitable inconsistency in methodology. This also highlights the critical constraint when it comes to benchmarking of practices and outcomes, as it has become relative.

Similarly, situation or tool finds prominence in the whole episode matters, but the methodology must not lean on the effectiveness of a technique rather on the appropriateness of the protocol and efficiency of its implementation. Therefore, there has to be a clarity in approach that will stabilize, steer, and succeed in achieving objective. Waste reduction is not an assurance on cost of quality nor productivity will decrease and gross profit will rise which will take the excellence professional to stardom [14]. A naïve implementation of Lean is a demonstration of few techniques and tools with absolute disregard to the problem at hand and waiting for a magical improvement to follow.

Lean is tightly coupled with production planning and control system; therefore, a naïve implementation will make inadequate changes in isolated regions that will not be sufficient to create an impact on the ecosystem and thus, the objective of organizational transformation will hit a logjam. Therefore, in a haste such a lean approach bend on tools that fails to collectively build on benefits and cover the optimization of the entire system leading to an incomplete appreciation of the role of leadership for organizational development. Then, such a Lean approach end up as a mean-based approach that cannot assure enduring benefits.

Argument 10: A method that creates a culture.

If Lean is a culture, then the cultural elements must be transient in the society and political forces must support it, then only a change management leadership that fosters the outcome can survive. The methodology, techniques, people empowerment, human relations, communication must work synergistically to create a milieu where continuous transformation as a culture thrives. Neglecting role of consensus and collaboration that forms the bedrock of human relations will jeopardize the outcome of the strategy [29]. Situational relevance and deliberate temporal progression as critical variables; with choice of tools and risk management as residuals determines the fate of the Lean implementation strategy. Only a learning organization from its mistakes, culturally tuned for waste elimination, and iterative implementation alone can churn value out of a system. Lean fails to have a phase and an analytical schema to connect factors such as intuitiveness, perception, judging etc. as human factors that contributes to errors, thus creating process prisoners created by our espoused past. In the effort to dehumanize processes, are they not antithetical to the claims on human resources competency and conveniently being sidelined in Lean when it comes to defects management? While people are not monitored, the people scrupulously monitor process with stopwatches and the numbers are only seen from the point of process disorders while process masters might leave unscathed. By any chance, if numbers are looked from a people angle, then it is invariably will find an overarching reason to settle scores and materialize individual agenda. Have that not made Lean a methodology averse to radical innovations? This is in addition to the failure mode created by the inadequate implementation of a successful strategy. By not having a Lean hierarchy to drive initiatives, the first among equals norm sets in, then it becomes persuasiveness of the phrases, placards and parades unleashed by power brokers in enterprises that determines the effectiveness of a methodology rather than the value it actually generated.

Argument 11: A method that makes difference by setting house in order.

When problem lies outside the production house, then how best a leaning drive inside the house bring betterment to overall process. For example, congested roads and supply ecosystem widely distributed, and their supply chains are not optimized will bring more anonymity and failures in the JIT process. The extended gap between Lean intervention storms also added to reasons to lose momentum and motivation.

Argument 12: A method with a human centric face.

The role and competency of Lean management leadership many times by the nature and style of intervention, and by the demands set on employees (long working hours, denial of leave, lack of support for their suggestions, ad-hoc project management, no lead-by-example, only demands) has only created unhappiness and loss of confidence. In an effort to build speed, management fails to realize that the same thrilling speed can even kill the quality and creativity in business. To support the rush, specialization becomes a must that takes away the variety and boredom sets in.

In a hurry to crunch the unproductive training time, create parallel training tracks but that denied integrated competencies and holistic visibility. Limiting the training to senior and middle segments of organizational hierarchy, companies make the mistake of leaving the larger section of the workers on the floor guessing on the developments and builds anxiety. The behavioral changes from anxiety to discontentment to rebellion is not a long way which will manifest as loss in productivity and quality which will defeat any process improvement irrespective of however better it may be.

Laxity in leadership commitment and failures in creating breakthrough in culture change management leads to failure of Lean management. Minor glitches during test and transition phase of any Lean life cycle, if it rakes up restlessness in leadership leading to issuing discomforting communications, adds fuel to failure. The cumulative effect is loss in trust and respect to the management. Eventual Robotic Process Automation decisions will lead to elimination of manpower and even necessity for further lean intervention. Why there was no analysis on the failure in responsibility and accountability of senior management in ensuring the culture change? Where lies the human centricity?

5. Outcome as recommendations -6Cs

5.1 Command

Lean cockpit must be responsible for selection and approval of Lean initiatives are as well considered as projects for an organization. Unless the definition and disciplines of the project are not mandated, these improvement initiatives will go never ending. Lean is a strategic service provided by the corporate improvement groups. Therefore, it is imperative that there must be a protocol for driving a Lean intervention and solution selection, and a structured approach to choosing the project. The project must have a scientifically estimated quantitative business impact upon which outcome of the project may be evaluated. The projects must cover process and product performance parameters independently. The project must cover processes for development and support of the products. Projects must ensure the products and processes are improved and innovated with Lean implementation. The collective nature of achievement of a business objective must be mapped and drafted as part of charter, then project specific expected contribution must be defined. A collective projects catalog may be transformed into a monitoring dashboard. Milestone based verification and validation needs to be performed to control execution of Lean drive.

5.2 Control

Leaning operation involves a thorough change management. Then a guarded approach to the culture change implementation is necessary. A change approval board must intervene to verify and validate proposed new working model, this is necessary for wider acceptability as is takes the ownership from a lean Leader to an institutionalized arrangement. Prior to change adoption, configuration management audit must be performed to ensure continuity, integrity, safety and security of the business. Any setback in the process of the Lean initiative must be recorded as a leaning incident and a cause and consequence assessment needs to be performed before fixing.

5.3 Culture and competency

Lean is ultimately a culture change promoted by people, process and technology. Out of which people forms the most critical link that determines the make or break of this initiative. It is the insecurity feeling that originates from fear of job loss, redundancy, anxiety over inability to perform in changed circumstances breeds collective objection that ends in obstruction of the change. But the whole series of fear psychosis is rooted in concerns around the incompetency factor. Therefore, creating the right milieu is essential as part of planning an enterprise wide Lean initiative. A thorough competency analysis that ends in a competency development plan is a must that takes care of the career aspirations and employability of the affected population, thus transforming them as Lean promoters. The learning and development imparted to the pioneer group may be systematically percolated to lower layers to enable organization wide understanding that takes away shocks and surprises.

5.4 Cause and consequence

A proactive extended Failure Modes and Effects Analysis that maps potential failures to their possible consequences. A systematic assessment of systemic failures in Lean is essential to ensure erosion is value is prevented and savings to be sustained. The major sources of failures observed in this study are management, competency, communication, leadership, teaming, performance measurement, suppliers. Every source has their own set of categories of failures. Management with their lackadaisical attitude, uncommitted sponsorship, under budgeting resources, and ill-informed as major categories contribute to failure. Under competency identifies, inadequate and ineffective learning and development programs in lean have been contributing to failures. Communication as another source of failure identifies weak and limited broadcast that lacks conviction. The leadership quality as a source is another concern, when process identification, project identification, ensuring cooperation and breakthroughs are identified as categories of failure. Teaming is another source of worry, when team displays disloyalty, non-committal, under involved, uncertain, incompetence as categories of failures. Performance of a project is another source of worry as data, measurement, analysis, reporting, action orientation are categories of concern. Suppliers are another significant source of failure, as categories such as poor-quality supplies, and delayed supplies are major categories that attribute to Lean initiative failures. All the categories may be subjected to intense independent Root-Cause Analysis from all the 7 M dimensions such as man, money, material, method, machine, measurements and milieu can reveal the ultimate root causes which when acted upon will help in arresting repeat failures. Collective analysis of these root causes enables organizations to plan preventive actions against potential failures.

5.5 Communication

Uniform understanding of the objective and approach among stakeholders is essential to get the buy-in from the enterprise to succeed in Lean initiative. The program charter, risk and repercussion analysis, performance measurement analytics are key information that needs messaging. External suppliers are critical stakeholders as any service quality parameter slippage may adversely impact the overall performance of the process despite rigorous leaning efforts internally.

5.6 Convergence

Convergence of purpose is essential to find business impact. Convergence of plan is essential to find focus of all values derived to a financial benefit. Convergence of skills, knowledge, abilities, aptitude and attitude is essential to translate the gains to tangible outcomes and sustain the same. This convergence can become a reality, if and only if, labor feels stabilized and confident. The work force must feel owned by the corporate, then they will own the transformation agenda.

6. Concluding thoughts

Organizations across globe in pursuit of competitive advantage has a history of having implemented a variety of policies and strategies to gain competitive advantage. However, all such attempts are not accompanied by successes, rather many face obstacles and failures in the implementation of these programs. Organizations invariably turn the blame on to staff commitment and performance. Confirming the same as applicable to successful implementation of lean programs and achieving the benefits depends on the quality, preparedness and readiness of the human capital. Therefore, the organizations should know, the men behind the machines. If employees are not adequately taken into confidence in the implementation of lean programs, the benefits will not accrue. This is because, mostly imperfections marred implementations in supplier ecosystem.

Recent years when there is a global rise in evangelism over lean with coherent and persuasive arguments, there are a very few self-conscious attempts to critique this methodology from academic community. The criticisms are never an assurance of value when adequately considered during lean implementation but if not done, then research community fails in its fundamental responsibility to ensure subjugation of philosophy to insidious academic scrutiny to loosen the grip of vested interests on a management dogma and subsequently bring down totalitarianism. Else, blind following of assurances of wizardry by a technique will reap disappointing performances forcing the gullible professionals taking cover behind another hype and move with the tide, but failure modes never comes obvious for an open debate. Choice of blending a cocktail of tools is part of building technical rationality, but if it disempowers the human factor, then it brings a crisis. Failure of Just-in-time, absence of safety stock, and field defects creates an opportunity cost attributable to the methodology. In the aggression to remain lean, even CSR and employee welfare spending may get treated as non-value adding. Management experimentations thrive where collective deliberations are minimally promoted, which means, democratic values in industrial relations are not of prime importance, rather it means trade unionism must fail. The fundamental axioms of enterprise unions such as healthy, safe, fearless, work environment that assures indiscriminate treatment and respects individuality of thoughts and deeds; and people are permanent assets are destined to remain in chapters. Constancy of workforce is a wild assumption in Lean.

By performing current research author has proved that if arguments mentioned above are taken into consideration and are actually managed then companies has all prerequisites to achieve its desired targets in terms of lean – meaning successful lean implementation. At last, creation of the lean as a value system is central part of the model; it drives all other steps and thus is main critical success factor for the successful lean thinking implementation. Thus, the initial proposal is true – on of the reasons of failure of lean thinking implementation process is absence of company's vision of its lean initiative in the form of lean strategy as a form of their unique lean value system.

The findings of this study are only the tip of the iceberg. There are a lot of questions which should be answered in this area. Even while failures are rampant, still corporate herd around Lean, what all could be the driving factors? How to study a Production System? Are these philosophies still relevant today in the times of robotic automation? Are there other ways to study lean system without visiting a site what will be the output of such a travel?

Another question which might arise is how to create own lean value system representation – where and how to start. And there are more such questions. To answer all of them the ultimate goal has to be achieved – Development of the curated model of successful lean implementation. This model should incorporate Process, People and technology aspects for all manufacturing process types facilitating assessment of the financial feasibility of implementation as well. Ultimately, the transformation must lead to lean capital needs and generous labor needs to assure and ensure an all inclusive social economic growth, which is part of corporate social responsibility which every corporate strategy has to support.

Author details

Sajit Jacob[1*] and Krishnamurthy Kothandaraman[2]

1 Bharathiar University, Coimbatore, India

2 Management Consultant, Chennai, India

*Address all correspondence to: sajitjacob@rediffmail.com

References

[1] Nachiappan, R., and N. Anantharaman, 2006, "Evaluation of overall line effectiveness (OLE) in a continuous product line manufacturing system", Journal of Manufacturing Technology Management, Vol.17(7), pp.987-1008

[2] Bayat, H., and M. Dadashzadeh, 2017, "The impact of organizational factors on implementation outcomes of lean manufacturing", Journal of Business and Economics Research, Vol.15(2), pp.33-44.

[3] Worley, J. M. and T. L Doolen, 2006, "The role of communication and management support in a lean manufacturing implementation", Management Decision, Vol.44(2), pp. 228-245.

[4] Abdulmalek, F. A., and J. Rajgopal, 2007, "Analyzing the benefits of lean manufacturing and value stream mapping via simulation: A process sector case study", International Journal of Production Economics, Vol.107(1), pp.223-236.

[5] Jadhav, R.J., S. S Mantha, and B.S Rane, 2014, "Exploring barriers in lean implementation", International Journal of Lean Six Sigma, Vol.5 (2), pp. 122-148.

[6] Poksinska, B.,2010, "The current state of Lean implementation in health care: Literature review", Quality Management in Healthcare, Vol.19(4), pp.319-329

[7] Ahuja, I., and J. Khamba, 2008, "Strategies and success factors for overcoming challenges in TPM implementation in Indian manufacturing industry", Journal of Quality in Maintenance Engineering, Vol.14(2), pp.123-147.

[8] McKone, K. E., R.G. Schroeder, and K.O. Cua, 2001, "The impact of total productive maintenance practices on manufacturing performance", Journal of Operations Management, Vol.19(1), pp.39-58.

[9] Propate, R. V., S. R Jachak, and P.A Hatwalne, 2011, "Implementing approach of total productive maintenance in Indian industries & theoretical aspect: An overview", International Journal of Advanced Engineering Sciences and Technologies, Vol.6(2), pp.270-276

[10] Lodgaard, E., J. A. Ingvaldsen, S. Aschehoug, and I. Gamme, 2016, "Barriers to continuous improvement: perceptions of top managers, middle managers and workers", Procedia CIRP, Vol 41, pp. 1119-1124

[11] Zhang, L., B.E Narkhede, and A.P Chaple, 2017, "Evaluating lean manufacturing barriers: an interpretive process", Journal of Manufacturing Technology Management, Vol. 28(8), pp.1086-1114

[12] Bhasin, S.,2012, "An appropriate change strategy for lean success", Management Decision, Vol. 50 (3), pp. 439-458.

[13] Staudacher, A.P., and M. Tantardini, 2008, "Lean production implementation: a survey in Italy", Dirección y Organización, Vol.35, pp. 52-60.

[14] Rivera, L. and F. F.Chen, 2007, "Measuring the impact of Lean tools on the cost-time investment of a product using cost-time profiles", Robotics and Computer-Integrated Manufacturing, Vol.23(6), pp. 684-689

[15] Voss, C., and N. Tsikriktsis, et al., 2002, "Case research in operations management", International Journal of Operations and Production Management, Vol.22(2), pp.195-219.

[16] Karlsson C. and Åhlström P. , 1996, "Assessing changes towards lean production", *International Journal of Operations and Production Management*, Vol. 16(2), pp. 24-41

[17] Stone, K.B., 2012, "Four decades of lean: Systematic literature review", International Journal of Lean Six Sigma, Vol.3(2), pp.112-132

[18] Kotter, J. P., 2007, "Leading change: why transformation efforts fail", Harvard Business Review, Vol. 85(1), pp. 96-103.

[19] Arfmann, D and F.G.T Barbe, 2014, The Value of Lean in the Service Sector: A Critique of Theory & Practice, International Journal of Business and Social Science, Vol. 5(2), pp.1-7

[20] Pearce, A., and D. Pons, 2019, "Advancing lean management: The missing quantitative approach", Operations Research Perspectives, Vol.6 (C), pp.1-14

[21] Holweg, M., 2007, "The genealogy of lean production", Journal of Operations Management, Vol. 25(2), pp. 420-437

[22] Womack, J.P, and D. T Jones, 1997, "Lean thinking, banish waste and create wealth in your corporation", Journal of the Operational Research Society, Vol. 48(11), pp.1148

[23] Womack, J.P., D. T Jones, and R. Daniel, 1991, *The Machine That Changed the World : The Story of Lean Production*, Harper Perennial, ISBN 978-0060974176

[24] Pakdil F., and K.M Leonard, 2017, "Implementing and sustaining lean processes: the dilemma of societal culture effects", International Journal Production Research, Vol.55, pp.700-717

[25] Radnor, Z. and Johnston, R. (2013). Lean in UK Government: internal efficiency or customer service? , *Production Planning and Control*, Vol:24(10/11), pp. 903-915

[26] Teehan, R. and W. Tucker, 2010, "A simplified lean method to capture customer voice", International Journal of Quality and Service Sciences, Vol. 2(2), pp.175

[27] Bowen, D. E. and W. E. Youngdahl, 1998, "Lean" service: in defense of a production-line approach", International Journal of Service Industry Management, Vol.9(3), pp.207

[28] Fukuzawa, M., 2019, "Critique on the Lean Production System Research", Annals of Business Administrative Science, Vol. 18, pp. 85-101

[29] Hendry, C., 1996, "Understanding and creating whole organizational change through learning theory", Human Relations, Vol. 49(5), pp. 621-641.

Section 2

Lean Manufacturing Practices and Environment

Chapter 12

Lean Manufacturing towards Green Manufacturing Practices and Its Implementation in SME's

J.P. Rishi

Abstract

The manufacturing SMEs are facing the burden of non-equilibrium of the supply–demand chain along with the global change in the climate. Several SMEs are looking for a substitute that can create a balance between performance and the environment. In spite of numerous studies related to green and lean that has been evolved, none of them is able to clearly define the spheres of green and lean. Here in this chapter, there is an exploration of advancement of lean and green manufacturing and its impact on other sectors. It also highlights the methodology adopted in implementing the same. This chapter recognizes the commonalities between lean and green approaches, the collaboration and impact, techniques involved. Also, the impediments and perplexities confronted by the manufacturing sector are examined. Further, this gives a better understanding of the challenges before implementing lean with green. This chapter also recognizes possible gaps in the literature that will help to eliminate the barrier toward this Neo manufacturing.

Keywords: Lean, Green, Manufacturing, Operation, Management

1. Introduction

Liberalization Privatization Globalization (LPG) has created market possibilities for manufacturing businesses around the Globe. This encourages competition worldwide for quality products, thereby leading to different ways to enhance their competency. A global phenomenon like financial instability, political imbalance, technology revolution, stringent regulation and global climate changes are affecting the competition [1]. Both energy and resources cost are always escalating due to non-equilibrium between supply and demand this is effecting the operation cost and economics of manufacturing directly. One vital strategy for SMEs to face competition to diminish the waste of resources by inefficient operation and production methods.

Numerous big Industries have implemented the lean-to step up competition worldwide [1]. This idea of lean is exceptionally well known and embraced in numerous businesses including Toyota, Boeing and Ford. With numerous elucidations for defining lean that have come to perplexity to recognize its effect. The description of lean itself is tricky. Some expressed lean has philosophy than a plethora of tools. On the other side, a viable and system approach depicts lean is an assortment of waste diminishment approaches. Also lean is a framework with the primary objective to dispense waste. In any case, the common lean plan centres on the diminishment of waste. The lean tactics offer technique, tools, practices, and methodologies which can be useful to the manufacturing framework resulting in fewer resources utilization and fewer wastes. Lean manufacturing is actualized within manufacturing facilities since the effective usage by the Toyota Production System (TPS). Lean Manufacturing is characterized as an approach that disposes of all non-value added activities in production. Whereas lean centres the operational part, green looks into environmental perspective. Green addresses ecological balance and includes distinctive apprehensions like a waste, air, water and land pollution along with recycling management to enhance productivity [2]. The European Commission expressed the principles and practice of green in liaison with other buffer zones thus creating green economy which enhances the product value with lesser utilization of environmental-friendly resources everyplace conceivable. The green economy is an imperative way to propel financial development in the advancement of reducing carbon emission in process and products. From the green development point of view, the World Bank (2012) characterized green growth as a development effective in utilizing normal resources, limit contamination and natural impacts and resilient towards normal risks. Green development underscores worldwide maintainability that empowers the world's poorest and most powerless to take advantage of proficient, clean and buoyant progress. Green also additionally branded as the modern Lean wherein businesses have begun to provide support [3]. It is proved that green component primarily centres on cutting down contamination by increasing the value using resources judiciously [4, 5].

Green manufacturing has numerous acknowledgements in the literature study. Green Manufacturing is a trading technique that centres on productivity by the responsive and active environmentally friendly process. It is also seen as a rejuvenation of the manufacturing sector [6]. It alludes to any Manufacturing unit which employs innovation and renewable sources of energy whereas "Green" refers to activity to decrease contamination and waste by the reduction in the resource. The study reveals that industries that have implemented innovative green management, not only progress but also fulfil the requests of the buyer to enhance reputation among controllers and the public [7].

The collaboration within lean and green will encourage and improve performance by creating valuable products and waste diminishment. This combination has blended and intrigued many analysts. Numerous rudiments drive the accomplishment of Lean Green Manufacturing are examined within this chapter. Its usage is a major challenge due to the want of a Lean green domain expert, thereby deferring its execution and decrease in industry competitiveness. This chapter surveys standard of Lean Green manufacturing and its execution strategies. Also covers their metrics and determinants, challenges and future prospectus about Lean Green manufacturing.

2. Lean Manufacturing

2.1 Introduction

Lean Manufacturing considered an important component of modern manufacturing units, subsequent to industrial revolutions. It removes activities/events that don't add to the product value. **Figure 1** depicts the milestones of Lean Manufacturing. With the onset of first industrial revolution in the UK in early 1800

1800	1890	1900	1910	1950	1990
E. Whitney	F. W. Taylor	F. B. Gilbreth, L. M. Gilbreth.	H. Ford	T. Ohno	J. P. Womack, D. T. Jones, D. Roos.
Interchangeable Parts	Time Study	Motion Study & Chart	Flow Process and Mass Production	Toyota Production System	Lean Manufacturing

Figure 1.
The road map of Lean Manufacturing post-Industrial Revolution.

by Eli Whitney where machines were for the first time used in manufacturing. He conceived the idea of interchangeability which permitted a huge number of unskilled labours to be employed in musket manufacturing which was considered as a skilful job done by craftsman only during that time. This substantially enhanced the scale of economics in the production units. The manufacturing units intensively relied upon specific technologies till the late 1890's when one F.W. Taylor brought time study to reduce processing time. This approach measures and examines the quantum of time essential for an operator to finish the given job by means of a stopwatch thus establishing "Standard time" to complete a definite task. Subsequently, Process control chart by Frank and Lilian Gilbreth and followed by Motion Study during early 1900. The process control chart is drawn to give an actual picture of the manufacturing process and is used to take corrective and remedial actions to improve the process. The motion study is a methodology adopted to reduce unnecessary motions required to perform the task. Now both Motion study and time study are improved further and integrated and to be called method engineering. It assesses the engrossment of human beings with devices and guides human beings to effectively accomplish the given task. These innovations are the originators of waste elimination in terms of motion and idleness which are considered as forms of waste in Lean Manufacturing.

2.2 Lean Manufacturing practices, principles and tools

Progress of Lean Manufacturing technique has a positive effect on the overall performance of manufacturing SMEs. This evolution in manufacturing from a traditional to lean method makes the SMEs have good productivity and in a modest manner. Lean itself is a philosophy that encapsulates productivity. Any Lean manufacturing principles are based on five principles constituting a systematic framework to effect lean implementation as shown in **Table 1** and pictorially in **Figure 2**. These principles identify non-value added activities in a production process using a variety of lean tools [8].

In short Lean Manufacturing, methodology improves productivity by creating value and reduces 7 types of common wastes encountered in a production scenario shown in **Figure 3**. In order to easily implement Lean manufacturing effectively in SMEs, a variety of Lean tools are available to support SMEs in order to implement lean. A variety of Lean tools are available as shown in **Figure 5** and their selection plays a very important role in improving the moral of SMEs in order to effectively implement the Lean Manufacturing technique. Few lean tools at an outset look to be similar in the name like process mapping or value stream mapping. Also, These Lean tools can be adopted in any SMEs situated in any country with suitable

Lean Attributes	Functions	References
Value	To recognize the customer's need for the product. This determines the willingness of the customer to purchase the product/service.	BMA (2008);
Recognize Value Stream	It makes the production department comprehend the Product life cycle and recognize the process or stages that don't add value to a product.	
Ensure Smooth flow	In order to smoothen the value flow, identify and eliminate the non-value-added process. Thereby raw material flows interruptedly and the end-user receives the products smoothly.	
Implementation of Pull-Based Production	Here production is based on demand. Production will be scheduled only based upon receipt of the Client order. This system requires flexible manufacturing systems and an agile supply chain network.	Crawford (2016)
Continuous Improvement	Aim towards perfection constantly improving the process.	

Table 1.
Five principle functions of Lean.

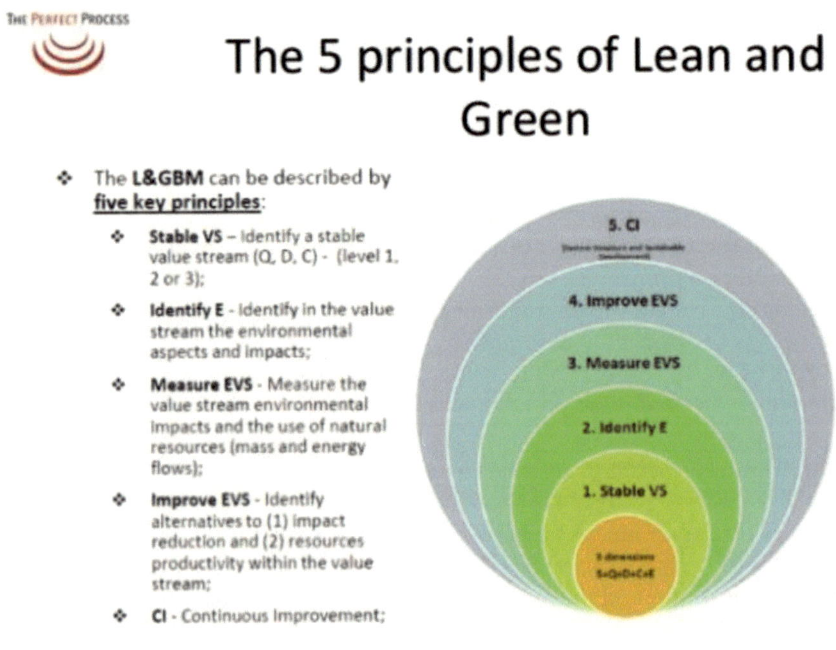

Figure 2.
The Five Principles of Lean and Green.

adjustments. Also, applying the wrong lean tools may cause additional resource consumption. In order to apply the correct lean tool, it is required to reclassify criteria for selecting lean Tool and map it with a particular type of waste that needs to be eliminated. So it requires a roadmap for implementing lean and it cannot be done overnight.

On successful implementation of Lean tools in SMEs resulted in enhanced Performance, Quality improvement, cycle time reduction. Lean tools enable smooth, well-organized layout yielding a high-quality product with low waste in SMEs Green factory shown in **Figure 4**. These tools have impacted significantly on manufacturing cost performance, waste reduction and improved performance [6].

Wastes	Descriptions	References
Over-production	Product that is produced without or beyond demand. Over-production is described as the main waste that will trigger and contribute to the remaining of 7 wastes.	Berg and Ohlsson(2005)
Waiting	Waiting for further product processing due to want of information and resources can cause waiting. Manpower is the main reason for waiting. Production flow interrupted is also considered has waste.	Bach (2017); McBride (2013).
Inventory	Inventory is a necessary evil in any manufacturing unit. High inventory results in locking up of cash flow. Also storage cost escalates.	Bach (2017); McBride (2013)
Motion	Motion is defined as an excessive movement of employee or machine that does not add value to the product Excessive motion might result from lack of standard operating procedure (SOP), poor layout design, etc. This will relate to employee health issue.	Abe (2015); McBride (2013)
Defect	A product that does not satisfy the end-users need. Defects can be resulted from poor quality control, poor inventory control or from machine breakdown.	
Transportation	Transportation is deemed as waste if it does not add value to the product, it might cause damage to the product or extra manpower to manage transportation. Transportation factor is normally caused by poorly designed layout, improvement can be done by simplifying the process and improve the layout.	Abe (2015)
Over-Processing	Over-processing is defined as manufacturing the product at the specification higher than customer's requirement.	

Figure 3.
Waste list according to Lean Manufacturing.

Figure 4.
Green factory.

2.3 Systematic implementation of Lean Manufacturing

Lean Manufacturing can improve existing productivity by eliminating non-value-added activity in the production chain. **Figure 5** enumerates vital lean tools used in any industry. Even though there are benefits with effective implementation of lean but still there will be few challenges left. Few lean tools incline towards similar intent of value making and they complement one another. For illustration, bottlenecking analysis, value stream mapping, continuous flow tools and Gemba analysis are used for analyzing block diagram for upgrading the value [9]. Many assortments of lean tools are available but due to poor guidance in mapping, the appropriate lean tool to beset non-value-added activity is tedious.

Normally Lean experts use a blend of different lean tools to formulate combined methods like Value stream mappings and kaizens. In the year 2015 a new model which integrate lean with the Green concept and Six Sigma using Define, Measure, Analyze, Improve, Control (DMAIC), which is parallelly implemented along with

Lean tools. To implement Lean Manufacturing the main basis is organizations culture and leadership to influence the follower's attitude and behaviour. Also, it is required to overcome inhibitors and identify critical success factors that are critical for implementing the integrated Lean Green model. Many studies discovered that human resources management is a very important critical success factor to implement successfully Lean Manufacturing in any organization. For this to happen a systematic top-down approach inside the SMEs is required [10].

3. Green Manufacturing

3.1 Introduction to Green Manufacturing

Global warming is escalating and people concerned are paying more attention to a clean and green environment. Even though Lean enhances performance in the

Top 25 Lean Tools

Lean Tool	What Is It?	How Does It Help?
5S	Organize the work area: • Sort (eliminate that which is not needed) • Set In Order (organize remaining items) • Shine (clean and inspect work area) • Standardize (write standards for above) • Sustain (regularly apply the standards)	Eliminates waste that results from a poorly organized work area (e.g. wasting time looking for a tool).
Andon	Visual feedback system for the plant floor that indicates production status, alerts when assistance is needed, and empowers operators to stop the production process.	Acts as a real-time communication tool for the plant floor that brings immediate attention to problems as they occur – so they can be instantly addressed.
Bottleneck Analysis	Identify which part of the manufacturing process limits the overall throughput and improve the performance of that part of the process.	Improves throughput by strengthening the weakest link in the manufacturing process.
Continuous Flow	Manufacturing where work-in-process smoothly flows through production with minimal (or no) buffers between steps of the manufacturing process.	Eliminates many forms of waste (e.g. inventory, waiting time, and transport).
Gemba (The Real Place)	A philosophy that reminds us to get out of our offices and spend time on the plant floor – the place where real action occurs.	Promotes a deep and thorough understanding of real-world manufacturing issues – by first-hand observation and by talking with plant floor employees.
Heijunka (Level Scheduling)	A form of production scheduling that purposely manufactures in much smaller batches by sequencing (mixing) product variants within the same process.	Reduces lead times (since each product or variant is manufactured more frequently) and inventory (since batches are smaller).
Hoshin Kanri (Policy Deployment)	Align the goals of the company (Strategy), with the plans of middle management (Tactics) and the work performed on the plant floor (Action).	Ensures that progress towards strategic goals is consistent and thorough – eliminating the waste that comes from poor communication and inconsistent direction.
Jidoka (Autonomatio n)	Design equipment to partially automate the manufacturing process (partial automation is typically much less expensive than full automation) and to automatically stop when defects are detected.	After Jidoka, workers can frequently monitor multiple stations (reducing labor costs) and many quality issues can be detected immediately (improving quality).
Just-In-Time (JIT)	Pull parts through production based on customer demand instead of pushing parts through production based on projected demand. Relies on many lean tools, such as Continuous Flow, Heijunka, Kanban, Standardized Work and Takt Time.	Highly effective in reducing inventory levels. Improves cash flow and reduces space requirements.
Kaizen (Continuous Improvement)	A strategy where employees work together proactively to achieve regular, incremental improvements in the manufacturing process.	Combines the collective talents of a company to create an engine for continually eliminating waste from manufacturing processes.
Kanban (Pull System)	A method of regulating the flow of goods both within the factory and with outside suppliers and customers. Based on automatic replenishment through signal cards that indicate when more goods are needed.	Eliminates waste from inventory and overproduction. Can eliminate the need for physical inventories (instead relying on signal cards to indicate when more goods need to be ordered).
KPI (Key Performance Indicator)	Metrics designed to track and encourage progress towards critical goals of the organization. Strongly promoted KPIs can be extremely powerful drivers of behavior – so it is important to carefully select KPIs that will drive desired behavior.	The best manufacturing KPIs: • Are aligned with top-level strategic goals (thus helping to achieve those goals) • Are effective at exposing and quantifying waste (OEE is a good example) • Are readily influenced by plant floor employees (so they can drive results)
Muda (Waste)	Anything in the manufacturing process that does not add value from the customer's perspective.	Eliminating muda (waste) is the primary focus of lean manufacturing.

Lean Tool	What Is It?	How Does It Help?
Overall Equipment Effectiveness (OEE)	Framework for measuring productivity loss for a given manufacturing process. Three categories of loss are tracked: • Availability (e.g. down time) • Performance (e.g. slow cycles) • Quality (e.g. rejects)	Provides a benchmark/baseline and a means to track progress in eliminating waste from a manufacturing process. 100% OEE means perfect production (manufacturing only good parts, as fast as possible, with no downtime).
PDCA (Plan, Do, Check, Act)	Iterative methodology for implementing improvements: • Plan (establish plan and expected results) • Do (implement plan) • Check (verify expected results achieved) • Act (review and assess; do it again)	Applies a scientific approach to making improvements: • Plan (develop a hypothesis) • Do (run experiment) • Check (evaluate results) • Act (refine your experiment; try again)
Poka-Yoke (Error Proofing)	Design error detection and prevention into production processes with the goal of achieving zero defects.	It is difficult (and expensive) to find all defects through inspection, and correcting defects typically gets significantly more expensive at each stage of production.
Root Cause Analysis	A problem solving methodology that focuses on resolving the underlying problem instead of applying quick fixes that only treat immediate symptoms of the problem. A common approach is to ask why five times – each time moving a step closer to discovering the true underlying problem.	Helps to ensure that a problem is truly eliminated by applying corrective action to the "root cause" of the problem.
Single Minute Exchange of Die (SMED)	Reduce setup (changeover) time to less than 10 minutes. Techniques include: • Convert setup steps to be external (performed while the process is running) • Simplify internal setup (e.g. replace bolts with knobs and levers) • Eliminate non-essential operations • Create standardized work instructions	Enables manufacturing in smaller lots, reduces inventory, and improves customer responsiveness.
Six Big Losses	Six categories of productivity loss that are almost universally experienced in manufacturing: • Breakdowns • Setup/Adjustments • Small Stops • Reduced Speed • Startup Rejects • Production Rejects	Provides a framework for attacking the most common causes of waste in manufacturing.
SMART Goals	Goals that are: Specific, Measurable, Attainable, Relevant, and Time-Specific.	Helps to ensure that goals are effective.
Standardized Work	Documented procedures for manufacturing that capture best practices (including the time to complete each task). Must be "living" documentation that is easy to change.	Eliminates waste by consistently applying best practices. Forms a baseline for future improvement activities.
Takt Time	The pace of production (e.g. manufacturing one piece every 34 seconds) that aligns production with customer demand. Calculated as Planned Production Time / Customer Demand.	Provides a simple, consistent and intuitive method of pacing production. Is easily extended to provide an efficiency goal for the plant floor (Actual Pieces / Target Pieces).
Total Productive Maintenance (TPM)	A holistic approach to maintenance that focuses on proactive and preventative maintenance to maximize the operational time of equipment. TPM blurs the distinction between maintenance and production by placing a strong emphasis on empowering operators to help maintain their equipment.	Creates a shared responsibility for equipment that encourages greater involvement by plant floor workers. In the right environment this can be very effective in improving productivity (increasing up time, reducing cycle times, and eliminating defects).
Value Stream Mapping	A tool used to visually map the flow of production. Shows the current and future state of processes in a way that highlights opportunities for improvement.	Exposes waste in the current processes and provides a roadmap for improvement through the future state.
Visual Factory	Visual indicators, displays and controls used throughout manufacturing plants to improve communication of information.	Makes the state and condition of manufacturing processes easily accessible and very clear – to everyone.

Figure 5.
Top 25 Lean Tools with their applications.

SMEs but does not consider any environmental factors during its implementation. Green Manufacturing is popularly known as sustainable manufacturing, it is a business strategy yielding profit through proactive environmental friendly operating methods.

Green Manufacturing originated at the beginning of the 1990s by eco-innovation is a novel production method leading to mitigating environmental pollution and negative effects of resources utilized. Numerous research work has been done to discover the green approach in SMEs. It is considered as an environmental and economical-driven method to minimalize waste by designing an efficient process and selecting materials without compromising the environment factors. Only Best practice should be adopted so that Green Manufacturing becomes effective. The International Standard Organization (ISO) came up with an international standard for Environmental Management System known as ISO14001 during the year 2013. These standards act like a systematic continuous improvement tool to implement Green Manufacturing.

3.2 Green Manufacturing Philosophies

In order to accomplish Green Manufacturing, there is a prerequisite to set up a suitable method to achieve it. For this, there are one dozen Principles used as strategies for designing a green manufacturing process as depicted in **Figure 6**. Also, there is a need for assessment of product lifecycle, types of Energy and material used. After further developments, a basic framework consisting of five principles of Green Manufacturing was evolved as shown in **Figure 7**. The objectivity of both approaches highlights assessment in a manufacturing SME considering Product life cycle (PLC), by offering deep importance to eco-friendly resources [11].

No.	Definitions	Descriptions
Principle 1	Non-hazardous inherent rather than circumstantial.	Strive to ensure all material and energy input and output are inherently non-hazardous as possible.
Principle 2	Prevention instead of treatment.	Better to prevent waste than treat or clean up waste after it is formed. Waste correlates to a process or an energy that is not being used efficiently in an operation.
Principle 3	Design for separation.	Product separation and purification process in manufacturing process consume a large amount of energy and material. Minimum energy consumption and material use design consideration should be incorporated in separation and purification operation.
Principle 4	Maximize mass, energy, space and time efficiencies.	A process is considered not efficient when mass, energy, space and time are utilized below maximum efficiency. Space and time can be utilized along with mass and energy to eliminate waste. Product, process and system should be designed to maximize mass, energy, space and time efficiencies.
Principle 5	Output-pull.	The production is based on client demand where material or energy will only be consumed when there is a demand.
Principle 6	Conserve complexity.	High complexity should correspond to reuse. The material used should have the complexity benefit without the need of modifying existing manufacturing process. Embedded entropy and complexity must be viewed as an investment when making design choices on recycle, reuse, or beneficial disposition.
Principle 7	Durability rather than immortality.	A product that lasts well beyond its commercial life will normally cause an environmental problem. Product design should consider the expected lifespan of the product. Design goal should be targeted at durability, not immortality.
Principle 8	Meet need, minimize excess.	The operation cost for material and energy cost can be high for an overdesign system. Design for unnecessary capacity or capability should be considered as a design flaw.
Principle 9	Minimize material diversity.	Material diversity in multicomponent products should be minimized to promote disassembly and value retention.
Principle 10	Integrate local material and energy flow.	This focuses on heat and material recovery on existing process. Design of products, processes, and systems must include integration and interconnectivity with available energy and materials flows.
Principle 11	Design for commercial "afterlife".	The recycling element should be incorporated into product design. This enables the current product to be utilized.
Principle 12	Renewable rather than depleting.	Taking waste product from a process and utilised as a feedstock into other process is a recycle/renewable source. The renewable resources can be utilized for recycling. Material and energy inputs should be renewable rather than depleting.

Figure 6.
Green Principles and their scope.

No.	Definitions	Descriptions
Principle 1	A comprehensive systems approach must be used to evaluate and improve manufacturing process from a green perspective.	Principle 1 takes environmental impact into consideration as the impact may have resulted from the process itself.
Principle 2	The system should be wholly viewed across both the vertical and horizontal directions.	Vertical refers to system varying level of detail from the enterprise down to the process while horizontal refers to the system at the same level of detail. This approach is important as environmental impact can occur depending on the level of detail.
Principle 3	Harmful input and output of the system to the environment and humans should be reduced or removed.	Identify harmful input and output to be replaced with material that has a lower impact. An alternative way is to implement recycle, reuse and remanufacture to reduce harmful input required.
Principle 4	Net resource use should be lowered.	Ideally, net resources should be zero so that resources may be used at a rate equal to the rate of replenishment in the environment. Aside from being practically impossible, the system will be deemed as inefficient in minimizing environmental impact.
Principle 5	Temporal effect on the system should always be considered.	Temporal effect is referred to effect that is time-based. Environmental impact should be considered in the design stage to allow better ability to consider for an alternative in reducing the future effect.

Figure 7.
Refined Green Principles.

The product lifecycle (PLC) rudiments viz., designing, procuring, manufacturing, package and delivery, old product disposal and reuse, need to have Green Manufacturing ingredients. After extending the further analysis of product lifecycle stages, a more refined framework is evolved by polishing principles, practice and tools involved in lifecycle design. The proposed framework has three main segments viz., design, development, manufacturing and services. This framework emphasis on resource utilization, Production Planning and Control (PPC), product assembly and warehousing, packaging. A environmental-based study reveals that about 80% impact on social, economic and environmental is found out in the design of the Product/Process. Therefore a new study Design for Environment (DfE) is presented which includes elements of green technology in designing process/product based on its impact on the Product life cycle. One more analysis called life cycle assessment (LCA) on a product is comparatively vital in the first stage of product design by analyzing key environmental factors that influence the complete product life cycle [12]. These two analyses are closely related to emission/waste and are important in assessing the product features leading to environmental pollution.

In a Product Along with DfE and LCA, energy consumption pattern plays a vital role in SMEs. Nowadays Green Ideology is no longer a new word to the entire world. It is defined as the "usage of science in the environment to preserve the resource and environment to control the negative impact of human activity". Few Energy-efficient technologies viz., co-generation, photovoltaic, biogas, etc. can enhance the green quotient of any SMEs. Even though settled green technologies are available they are not utilized to their full potential. So a lot of changes are required in terms of technology and knowledge in order to achieve complete Green Manufacturing. Screening of new technologies to lessen pollution is a must-have a positive effect on the Management of Energy by optimization of resources in SMEs. Example Gas Turbine (GT) co-generation plant can cut operation cost by developing two or more forms of energy from a single source [13].

A Material requirement planning system (MRP) should consider material identification, selection procurement and consumption to help in understanding both quantitative and financial term. Also, an analytical model to assess and analyze the effect of environmental risk factors during manufacturing which can produce different types of waste. Industrial collaboration lets SMEs acquire quality resources having low waste deposition. This type of transformation process will involve adaptation, replacement and circumvention steps.

- The adaptation steps involve collecting waste material from a collaborators area and fed it as an energy source.
- The replacement steps will replace stock by adjusting the alternate stock.
- The circumvention step stresses reducing/eliminating waste generation that may cause environmental pollution.

3.3 Enablers of Green Manufacturing

Green Manufacturing as similar waste reduction resourcefulness as Lean Manufacturing. So it is important to bring into line green and lean waste in order to bridge the void between both approaches and also correlate them to improve performance. Typical green wastes like greenhouse gases, high resource consumption like energy, water etc., pollutant, rubbish, and Safety and health [13].

An interview conducted with SMEs in Indian Industrial clusters (i.e. Foundry, garments and printing press etc.) on green technology and it was found out impediments for implementing Lean and green approach. The productivity of the workforce and lack of Lean and Green expert in SMEs results from sluggish implementations of Lean and Green. Also, Low or No cost improvement in SMEs is always welcomed because the management of these SMEs are not keen on huge investment in new technology, since they don't possess the required expertise and proficient employee to lead the team. Also return on investment is also important to convince the management. If SMEs focus on smaller energy-efficient equipment it can pay substantially towards overall growth. One main observation in SMEs is if any wing in a manufacturing SMEs if initiated with green technology will go on to complete the implementation.

According to a survey on the Energy consumption pattern by International Energy Outlook (2016), it was found that industry alone is consuming 55% of the world energy, and consumption is increasing by 2.1% in 2018. This Energy production and consumption of fossil fuel is the main reason for global warming. The main factors which can control this undesirable consumption rate are the legislature, corporate image, competition, and tax holiday. Also, stakeholders of SMEs can create an urge towards Green Manufacturing. Government play a pivotal role in encouraging and backing the SMEs in implementing Green Manufacturing. To achieve these mandate, the government has offered tax holidays for SMEs who have gone green and funding's and the incentive is given to Green SMEs will augment Green Manufacturing. One such case is of the Japanese government, which have shown commitment and zeal in realizing the green approach by introducing the Joint Crediting Mechanism (JCM) to have low-carbon technology collaboration with few developing countries. In return, Japan will have reduced total Green House Gas (GHG) emissions. Green technology like cogeneration/tri-generation can step up energy generation capability.

Today, educated customers have more awareness of global climate change and they can put pressure on the SMEs to include Green Manufacturing technology in their units. The customers are often pressurized and demand manufacturer which is directly related to competitor pressure. Many companies have improved their image and public perception by adopting green manufacturing technologies through their corporate social responsibility. Along with ISO 9000 certification for design and manufacturing company have to adopt green manufacturing technologies in order to acquire ISO14000 certification related to an environmental management system. ISO 14001 certified companies assess their supplier's environmental performances and compel their suppliers to adopt environmental practices [14].

Research unveils that Green Manufacturing mainly focuses on pollution, energy and waste management. Many SMEs around the globe have prioritized Green Manufacturing technologies and subsidy/incentives are given for their successful implementation. In facts, some SMEs have started green in their organizations without comprehending its actual benefits.

4. Lean integrated with Green Manufacturing

The objective of both the approach is same i.e. minimizing non-value-added activities they can very effective when unified and implemented simultaneously. The growth of Lean and Green manufacturing approach is recently initiated and not more than the past decade mainly both by researchers and Industrialist as shown in **Figure 8**.

Both Lean and Green Manufacture are similar and effective approaches that can improvise global competition by improving productivity with zero-waste. Various apps have been developed for Lean Manufacturing that can support SMEs to integrate smart systems in their up and downstream processes thereby completely reducing waste and later it Green Manufacture can be accomplished with a positive vibe on operational and environmental performances.

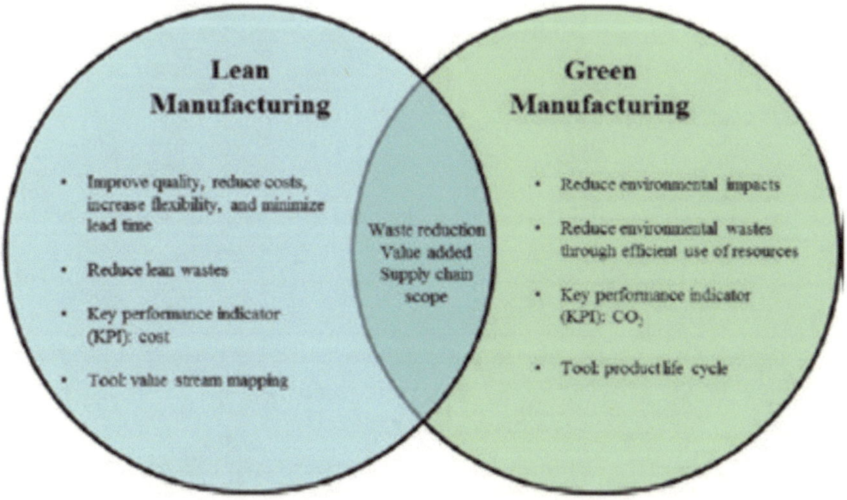

Figure 8.
Commonalities between Green and Lean Manufacturing.

Research shows that Lean Manufacturing creates a positive impact on the environment because these lean SMEs possess the ability to cut environmental pollution. SMEs who have adopted Lean Manufacturing for continuous improvement in their facilities resulted in high efficiency and better environment-friendly performance. Especially lean tools can be used to achieve Green Manufacture.

Implementation of Green Manufacturing has impacted positively on many SMEs by allowing the production utilities to perform better. It develops compatibility between environment and operation Management. The lean and Green approach is mutually complementary and augments the capability and efficacy of operations when related individually [15]. Both have the same objectives i.e. to enhance performance, quality, Lead time thereby reducing operation cost and creating a product with high values. This reveals the cross association between both manufacturing systems.

The integration of both approaches based on attributes like public and organization, waste, lead time, supply chain, tools, practices, performance indicators. **Figure 9** depicts the connection and lapping of Lean and Green, at an outset, both have the same traits but differently defined. The main similarity is in their objectives of waste reduction. It is seen that lean can maximize profit by reducing cost while green reduces environmental hazards thereby maintain ecological balance. Even though in **Figure 3** different objectives are stated for the lean and green approach, the Lean approach guarantees effective resource utilization by eliminating waste while the green approach reduces environmental pollution. So it can be said that both having different waste elimination approach but the same waste is targeted. This illustrates the symbiosis relation between the two approaches.

4.1 Impediments in the path of Lean and Green Manufacturing

Various Lean and Green tools are developed for its implementation but there is no validation of these method/model for an application in **Figure 10**. Manufacturing and Industrial engineering attract the attention of researchers in this area's

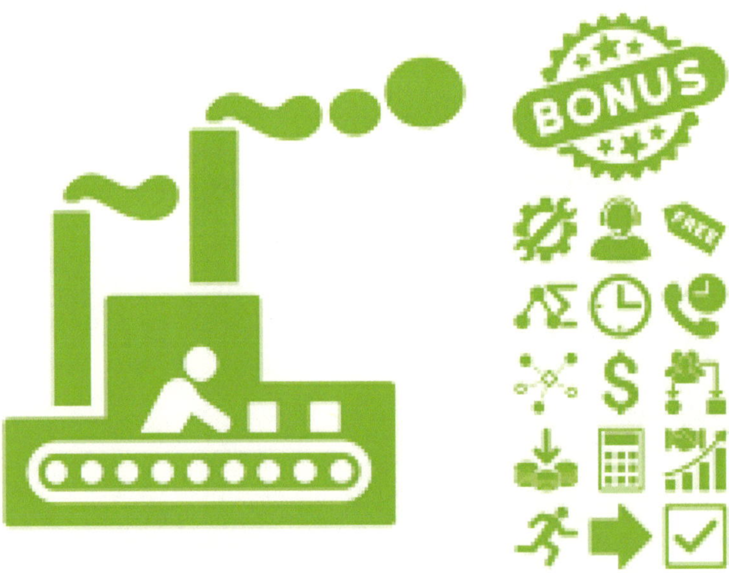

Figure 9.
Opportunities for Green manufacturing.

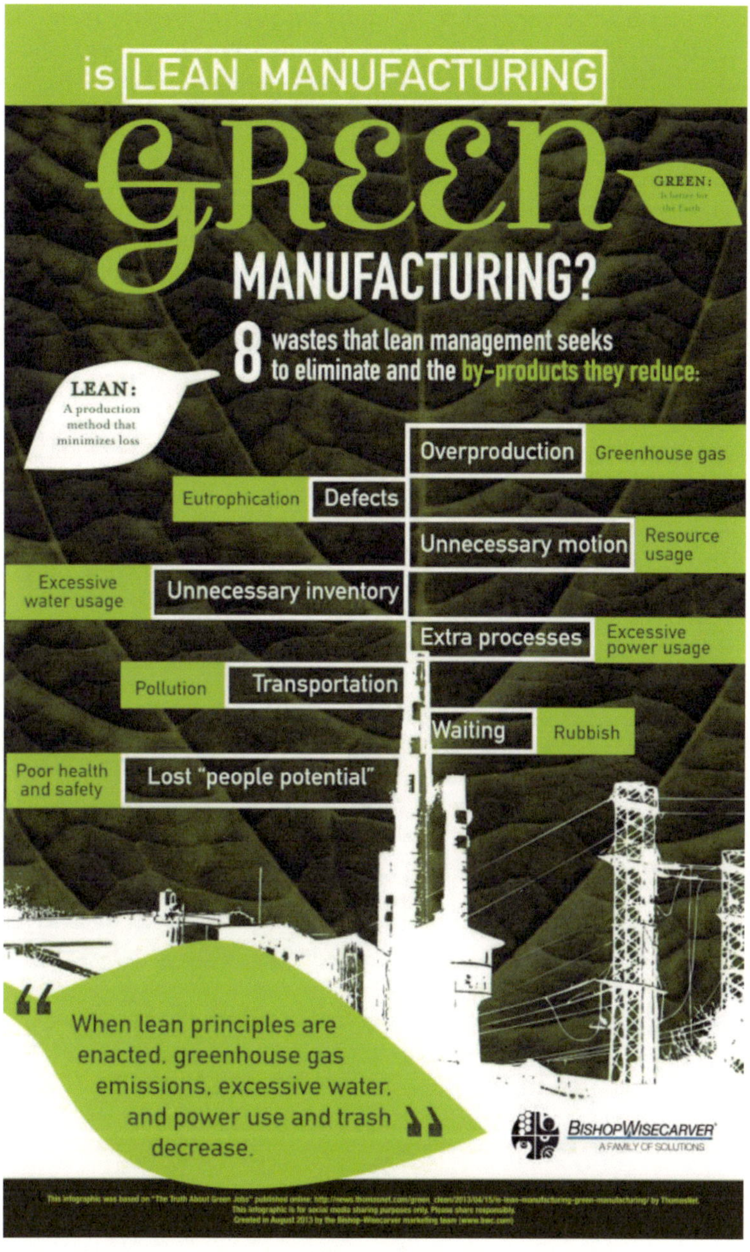

Figure 10.
Lean vs. Green.

leading to advancement. Also, management commitment is a key factor in implementing Lean and Green manufacturing and wherever there is a lack of organizational responsibility it becomes the main impediments for its implementation. Using the Management strategy and direction, the Lean and Green approach can be used as a continuous improvement program [16].

Lean and Green model can correlate waste data by using a systematic approach to support and realize d the benefit of this approach. Lean handles aids SMEs to transfer lean knowledge to the employee. From the operations perspective total productive maintenance (TPM) with root cause analysis are capable of smoothening production operation and from the green perspective, consumption can be controlled by optimization approach. By adopting energy management total operation cost can be cut off using energy saving opportunities. Environmental emission are continuously monitored and controlled by regulators in place. Today's SMEs is market demand driven, thus they need to adapt rapidly to latest technologies and performance boosters.

Figure 11 list trials faced while implementation of Lean green. The first step towards a green economy is the importance of endorsing teaching and training to the top management for their commitments towards implementing Lean and Green. For this Stake, holders are SMEs, NGOs, lean experts, Lean practitioners and academicians who can inculcate lean as well as green culture. Along with this, strong government backing is a must. Many SMEs are facing financial constraints in

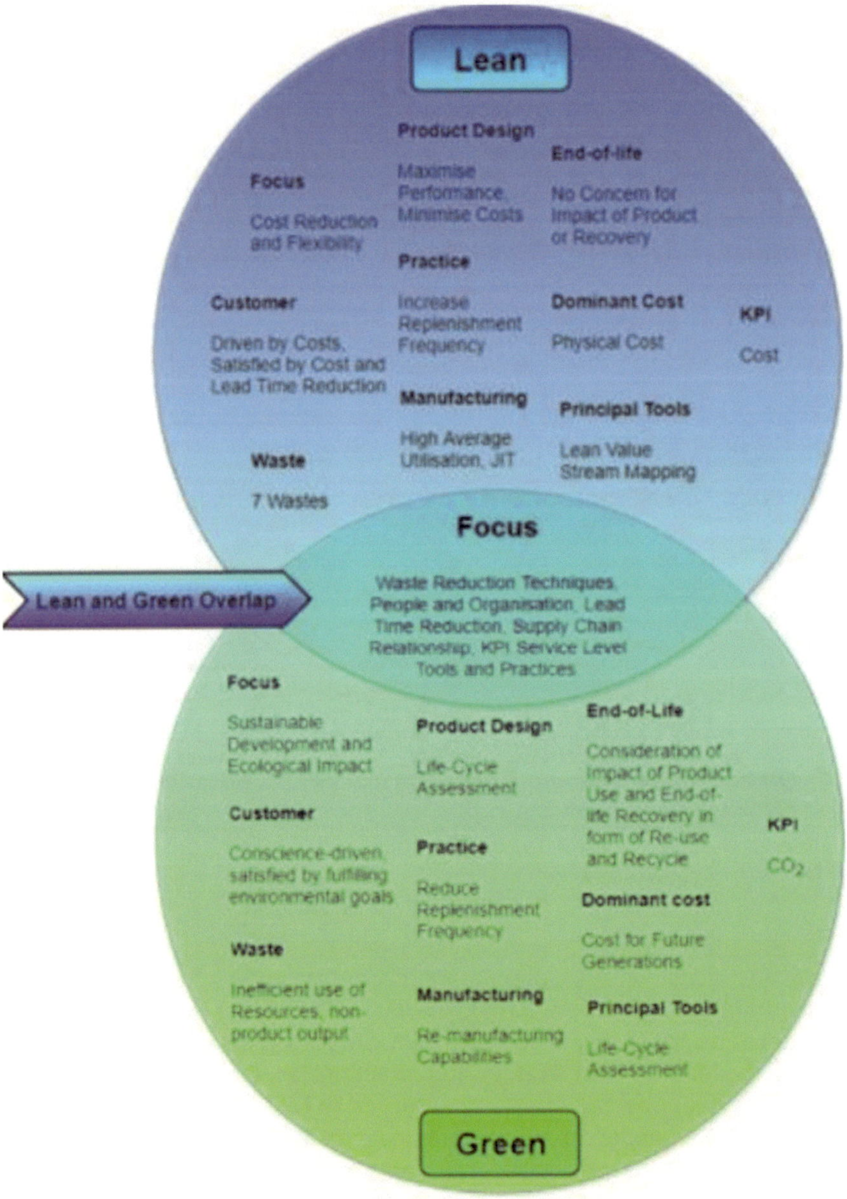

Figure 11.
Affiliation amid lean -green methods.

No.	Barriers in implementing L&G	Importance
1	Lack of environmental awareness	High
2	Fear of failure	High
3	Poor quality of human resources	High
4	Lack of expertise training and education	High
5	Fund constraints	High
6	Lack of statistical, lean and green thinking	High
7	Inappropriate identification of areas and activities to be Leaned and greened.	High
8	Lack of Kaizen culture	High
9	Lack of visual and statistical control during L&G implementation	High
10	Lack of government support to integrate green practices	High
11	High cost	High
12	Lack of communication and cooperation between departments	Moderate
13	Lack of top management involvement in adopting L&G initiative	High
14	Resistance to change	High
15	Poor corporate culture separating environmental and continuous improvement decisions	High
16	unreliable data collection system	

Figure 12.
Impediments of lean -green implementation.

implementing the Lean Green approach in their setup, so government policies play a vital role in facilitating funds through financial institution's. An important utility is the Management Information System (MIS) for the collection of Data and performance index, this enables SMEs to categorize problems and assess the efficacy of solution [17]. Also employing a capable individual to implement and progress Lean and Green is a big challenge. After reviewing the literature it is observed that there is no organized framework and analytical model to guide SMEs in refining their overall performances as shown in **Figure 12**.

5. Future work

The necessity to meet demands globally without harming the Environment.is the order of the day for many SMEs. Industry 4.0 posed challenges for SMEs in acquiring advanced technologies to be competitive. The digitalization of manufacturing has led the shift in industrial Engineering popularly recognized as industrial revolution 4.0 and resulted in smart factories, smart products and smart services embedded with the Internet of Thing (IoT). It integrates the production system with intelligence and creates new technology [18].

Today researchers have formulated guidelines for SMEs for implementing Lean and Green manufacturing. Collecting Data is the key to enhance SMEs performance. Using this data one can analyze operation parameters to step up productivity, quality and efficiency. To accomplish this SMEs need to embrace Industry 4.0, wherein the variety of smart sensors, expert system, advance solver, intelligent devices and data acquisition systems are a must. This requires huge investment by SMEs, who are not in the position to afford this upgradient. Therefore there exist a void in implementing Industry 4.0 with a confidence level within SMEs, with respect to the huge investment cost and global competition. To fill this void a model depicting flow in implementing Lean and green process of Industry 4.0 compliant is shown in **Figure 4**. This model can be implemented with confidence by SMEs at an

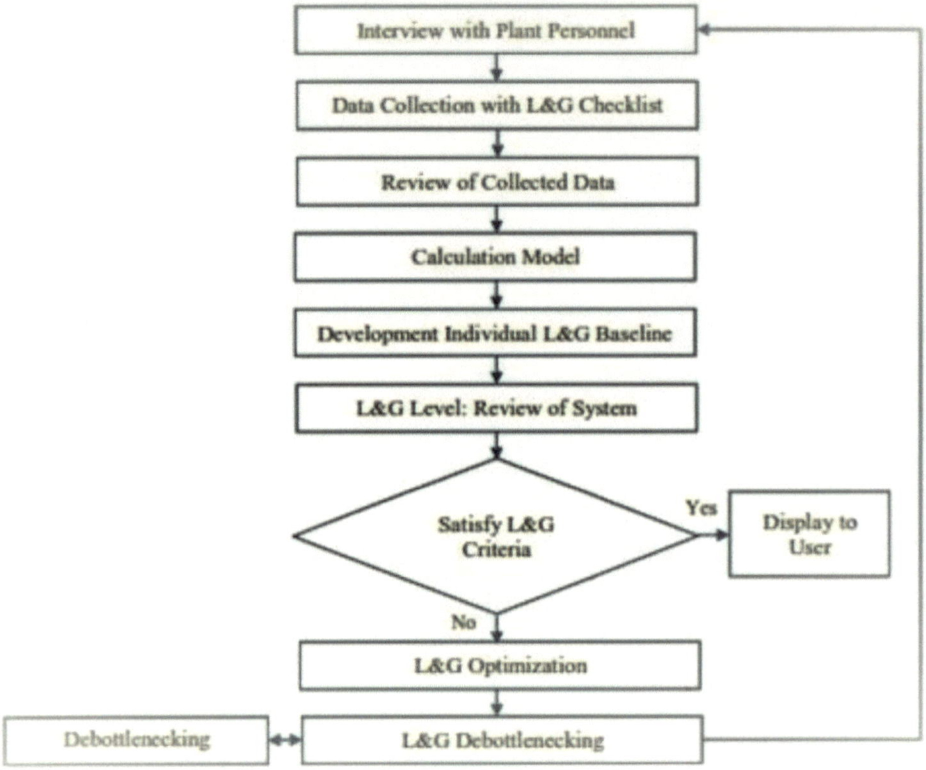

Figure 13.
Planned Process chart for Lean and Green.

affordable price. In order to adopt and implement Industry 4.0 an efficient data management system for further analysis permits continuous improvement.

A Lean and Green framework is proposed with collecting data and Lean and Green agenda as shown in **Figure 13**. This proposed framework primarily comprises five key elements such as manpower, money, machine, material and environment (5me). This proposed L&G framework alters 5me with the amalgamation of the green element. "Manpower" is the main element in SMEs, it has to adapt to the latest manufacturing technology, so a positive environment is a must for adapting rapidly. On the other hand, employee retention and capacity building are vital to

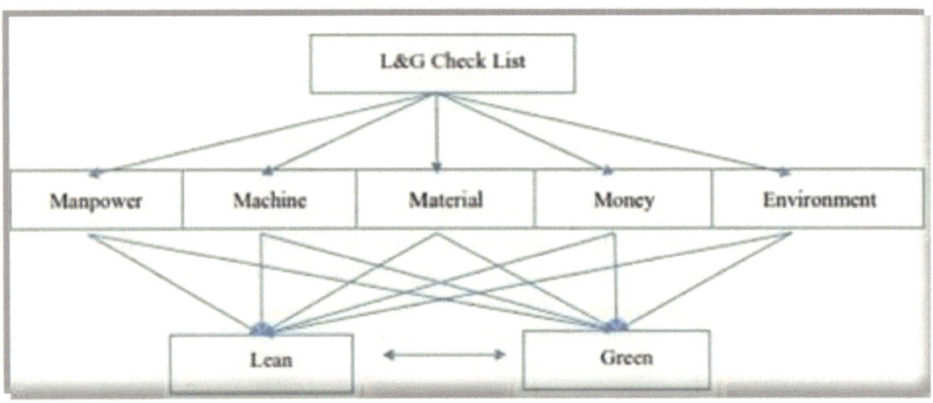

Figure 14.
Proposed Lean and Green manufacturing framework.

any SMEs functioning globally. There is relationship between cost-saving, efficiency and sustainability, the Money factor is a direct indicator of the performance index of an SME. Next is "machine", SMEs depends on a machine for manufacturing, so optimization of machine performance will improve production efficiency and create a continuous improvement environment for the employee to strive for better performance [19]. The element "material" includes resources like raw materials, products, logistic and storage, quality are considered. Finally, "environment" is involved to signify the green facet of the framework and it contributes severely to global warming and climate change.

These five major elements will help SMEs to implement the Lean and Green approach by giving direction regarding decision making by creating a multiple-criteria decision making (MCDM) tool which can help the organization to take the right decision based on their operational behaviour.

Also by using this proposed framework SMEs will be able to bridge the gap between an implementation using simulation technique used in implementing Industry 4.0. by combined development and analysis of process automation. The framework aims to overcome the barrier in implementing Lean and Green in the industry as shown in **Figure 14**.

6. Conclusion

The essentials of the Lean and Green facet has a foundation for beginners to comprehend the concepts of both is made. This integration leads to the synergy between Lean and Green yielding good result benefiting both environmental and operational performances. This chapter outlines the principles of Lean and Green manufacturing, tools for its application, enablers and inhibitors of this technology. The study reveals that due to ignorance, poor planning and no government policy to support and industries in implementing this approach. Total involvement of management within SMEs is a must for successful implementation. After identifying impediments of this integrated management still cannot implement effectively due to the shortage of experts in this domain. This chapter has reviewed the literature and identified the research gap amongst the barriers and implementation tools adopted, to fill this gap, a Lean Green framework model is developed to overcome the barriers of implementation. The ingredients of this proposed Lean and Green model are Men, Material, Machine, Money and Environment [20]. This model through simulation can fill the research gap to enhance the rate of implementation of this Lean and Green approach. This chapter can act as a primer to SMEs for a better understanding of Lean and Green, also to aid them to effectively implement sustainability in their manufacturing facilities. This chapter motivates and inspires SMEs to adopt Lean and Green manufacturing and to become Industry 4.0 compliant.

Abbreviations

DfE	Design of Environment
FPS	Ford Production System
GM	Green Manufacturing
IoT	Internet of Things
ISO	International Standard Organization
JCM	Joint Credit Mechanism
JIT	Just in Time

L&G	Lean and Green
LCA	Life Cycle Analysis
LM	Lean Manufacturing
MCDM	Multiple-Criteria Decision Making
NGO	Non-Governmental Organization
SMED	Single minute exchange die
TPS	Toyota Production System
UNEP	United Nations Environmental Program
WWII	World War II

Author details

J.P. Rishi
Department of Mechanical Engineering, VidyaVardhaka College of Engineering
Mysuru, Karnataka, India

*Address all correspondence to: rishijp@vvce.ac.in

References

[1] Anastas, P. and Zimmerman, J. (2003). Peer-Reviewed: Design Through the 12 Principles of Green Engineering. Environmental Science & Technology, 37(5), pp.94A-101A.

[2] BP.(2018). BP Statistical review of world energy. London: BP.

[3] Buesa, R. J. (2009). Adapting lean-to histology laboratories. Annals of Diagnostic Pathology,13(5), 322–333. doi:10.1016/j.anndiagpath.2009.06.005

[4] Business, B. (2018). 5 Ways How Globalization Impacts Small Businesses | BusinessBlogs Hub. [online] BusinessBlogs Hub. Available at: https://www.businessblogshub.com/2017/05/5-ways-how-globalization- impacts-small-businesses/ [Accessed 16 Jun. 2018].

[5] Carvalho, H., Azevedo, S. and Machado, V. (2010). Supply chain performance management: lean and green paradigms. International Journal of Business Performance and Supply Chain Modelling, 2(3/4), p.304.

[6] De Ron, A. and Rooda, J. (2006). OEEand equipment effectiveness: an evaluation. International Journal of . Production Research, 44(23), pp.4987–5003. Frank B, G. (2016). Process Charts. Wentworth Press.

[7] Galeazzo, A., Furlan, A. and Vinelli, A. (2014). Lean and green in action: interdependencies and performance of pollution prevention projects. Journal of Cleaner Production, 85, pp.191–200.

[8] Handfield, R., Walton, S., Seegers, L. and Melnyk, S. (1997). 'Green' value chain practices in the furniture industry. Journal of Operations Management, 15 (4), pp.293–315.

[9] Japan, G. (2017). Recent Development of the Joint Crediting Mechanism (JCM).Available at:https://www.carbon-markets.go.jp/document/20170522_JCM_goj_eng.pdf

[10] Jin Kim S. and Kara S. (2012). Impact of Technology on Product Life Cycle Design: Functional and Environmental Perspective. Berkeley. p. 1, 2

[11] Kochnev, I. (2007). WHAT, IF ANY, ARE THE DIFFERENCES BETWEEN THE TOYOTA PRODUCTION SYSTEM AND LEAN?. innovationlighthouse.com.

[12] McBride, D. (2013). The 7 Wastes in Manufacturing. Retrieved November 20, 2017, from a. http://www.emsstrategies.com/dm090203article2.html

[13] McGrath, W. (2007). Impact Analysis of Large Scale Lean Manufacturing Initiatives Upon Manufacturing a. Process Innovation In Irish Companies, Master thesis, Waterford Institute of Technology.

[14] Paul, I., Bhole, G. and Chaudhari, J. (2014). A Review on Green Manufacturing: It's Important, Methodology and its Application. Procedia Materials Science, 6, pp.1644–1649.

[15] Pettersen, J. (2009). Defining lean production: some conceptual and practical issues. The TQM Journal, 21 (2), pp.127–142.

[16] Qin, S. (2014). Process data analytics in the era of big data. AIChE Journal, 60(9), pp.3092–3100.

[17] Rouse, M. (2015). What is just-in-time manufacturing (JIT manufacturing)?. [online] WhatIs.com. Available at:

[18] a.https://whatis.techtarget.com/definition/just-in-time-manufacturing-JIT-manufacturing [Accessed 13 Jun.. 2018].

[19] Sezen, B. and Çankaya, S. (2013). Effects of Green Manufacturing and Eco-innovation on Sustainability Performance. Procedia - Social and Behavioral Sciences, 99, pp.154–163.

[20] TOYOTA. C. (2018). Toyota Global Site | Jidoka. [online] Toyota Motor Corporation Global Website. Available philosophy/toyota_production_system/ jidoka.html.

Chapter 13

Lean Manufacturing Practices and Environmental Performance

Ruhet Genç

Abstract

Lean manufacturing is considered a rewarding production strategy due to its positive effects on organizational and economic efficiency in various industries. Given the growing ecological consciousness, environmental achievements of lean manufacturing also incorporate a strong economic relevance. The main objective of this chapter is, therefore, to investigate the impact of lean manufacturing practices on environmental performance and the existing coherences between Lean and ecologically oriented variables such as resource usage, energy consumption, and air pollution. The methodology is literature review evaluating the findings of research in this sphere. Besides the discussion of its principles and methods, current trends and challenges regarding lean production as a business model that supports eco-efficiency are presented. The implications of this study will allow executives to better recognize and simultaneously solve both the economic and environmental problems posed by their companies.

Keywords: lean manufacturing, environmental impacts, eco-efficiency, lean, environmental performance

1. Introduction

Lean Manufacturing, also known as Lean Production [1], has long been considered an advantageous manufacturing method in various companies and industries on a global level as it aims for continually improving quality and efficiency in production process [2, 3]. It is essentially about eliminating the activities that do not add value in the production and seven types of waste, also referred as muda in lean settings [4] and includes several techniques such as Total Quality Management, Just-in-Time [5], Kanban, and Jidoka [6]. It has become essential with the increased price competition that companies experience in face of changing customer demands for goods and services by helping them produce same amounts at the lower costs.

In response to growing concerns for environmental issues since the early 1990s, stemming from the climate change and exhaustion of natural resources, the relationship between Lean and environmental performance has been a critical topic in business and manufacturing literature especially from 2008 onwards [7]. Moreover, numerous companies adopted lean measures to advance their manufacturing methods for decreasing their environmental burdens since Lean Manufacturing focuses on increasing production efficiency and decreasing waste.

Considering the higher efficiency provided by lean systems, which is the reduction in energy and material resources needed per unit output, the waste of production such as the emissions to air and water, generation of solid or hazardous waste, and chemical hazards abated along with the high costs of production. Thus, lean methods may alleviate environment by means of its inherent emphasis on waste reduction even though they do not automatically involve environmentally responsible intentions [8]. Research shows that although lean systems may not directly aim for reducing environmental effect, changing the production system to lean resulted in higher efficiency in energy consumption and in less waste [9].

The fact that incorporating lean manufacturing techniques may address environmental waste problem even without prioritizing it led companies to adopt lean activities. By integrating environmental considerations, lean manufacturing may provide businesses with long-term benefits such as environmental sustainability and recognition as a responsible corporate citizen. In this regard, companies embraced lean practices to reduce their ecological footprint for maintaining a better public opinion. Furthermore, many of them considered the "triple bottom line" of economic, environmental, and social concerns into consideration and took precautions on a bigger scale than those necessitated by the regulatory bodies [10].

The main objective of this study is therefore to provide insights into the contribution of lean manufacturing practices and measures for achieving a better environmental performance of manufacturing systems and the existing relationship of these two concepts. This is accomplished through a literature review along with a critical discussion of the findings. Focusing on the meaning of the green concept waste, lean manufacturing is described as a rising business model for endorsing environmental efficiency. The most effective lean practices and their tools in relation to supporting greener production are underlined. The improvement of the environmental measures of the firms that adopt the lean manufacturing principles, the reduction in ecological impacts stemming from their industrial operations are highlighted. The results of this study are of interest to scholars as well as to business managers to better understand and overcome environmental problems encountered by their business organizations.

2. Lean and green manufacturing: environmental impacts

2.1 Lean manufacturing concept and practices

Lean manufacturing techniques and methods, which provides higher efficiency than that its successor mass production processes, is based in Toyota Production System [2, 11] and the term 'Lean Production' was first coined by John F. Krafnick [12, 13]. To accomplish the aims like productivity, efficiency, profitability, product diversity, improved product quality, and customer satisfaction at higher levels [2, 14], the most vital practices of lean approach according to several scholars in the literature [5, 15–19], the tools they employ and their main contributions to production are presented in **Table 1**.

These practices, namely just-in-time (JIT), total productive maintenance (TPM), autonomation, value stream mapping (VSM) and kaizen or continuous improvement (CI), have been implemented by manufacturing businesses across the globe to realize the above-mentioned purposes of lean systems.

The five principles of Lean Production concept [1], displayed in **Figure 1**, are established to manage and minimize waste, which in turn protect the environment. Within the context of lean, waste which is considered any activity that does not add value to a product comes in seven main types and eighth types is included by Jeffery K. Liker [24]. These are in the form of overproduction, waiting, unnecessary transport, over processing, excess inventory, unnecessary movement, defects, and unused employee creativity as they are shown in **Table 2** along with their effects on environment.

Lean practice	Tools	Principle
Just-in-time (JIT)	Pull systems Takt time One piece flow Levelled production Cell manufacturing Visual control Kanban JIT purchasing Multifunctional employees	Reducing space utilisation, inventory cost and wastes that result from the overproduction
Total productive maintenance (TPM)	Single minute exchange of die (SMED) Overall equipment effectiveness (OEE) Planned maintenance 5S Quality maintenance Autonomous maintenance Initial control before starting production Safety and hygiene environment	Optimising predictive, preventive and corrective maintenance activities to achieve efficient and proficient production equipment
Autonomation/jidoka	Visual control systems A full working system mistake proofing devices	Reducing quality defects
Value stream mapping (VSM)	Flow diagrams Current and future state maps	Illustrating, identifying and measuring waste during the production process
Kaizen/continuous improvement (CI)	5S Continuous flow Run charts Five whys Brainstorming Data check sheet Kanban Pareto chart Gantt chart Mistake proofing Process maps VSM	Removing waste by improving operations gradually and continuously and sustaining lean system after its adoption

Table 1.
Most Essential Practices of Lean Manufacturing, adapted from [15, 16, 20–23].

2.2 Green lean concept and principles

Recently, carbon footprint and the environmental efficiency of the manufacturing companies has been an important topic [25] and green manufacturing concept whose goal is to continually incorporate better environmental conditions of manufacturing operations with the aims of mitigating air, water and land pollution that may pose threats to humans and other species and lessening the use of raw materials and energy, has emerged [26, 27]. It is defined as a series of activities that focuses identifying, measuring, evaluating, and managing the environmental waste created in various areas of production such as designing, manufacturing and planning [28].

Green manufacturing has both product and process perspectives. Concerning the product perspective, green manufacturing intends to produce eco-friendly products, keep the use of resources at a minimum level, and use materials that does not damage the nature. As for process perspective, it aims for minimum consumption of raw materials and energy, minimum dispersion of hazardous substances and minimum generation of waste [29].

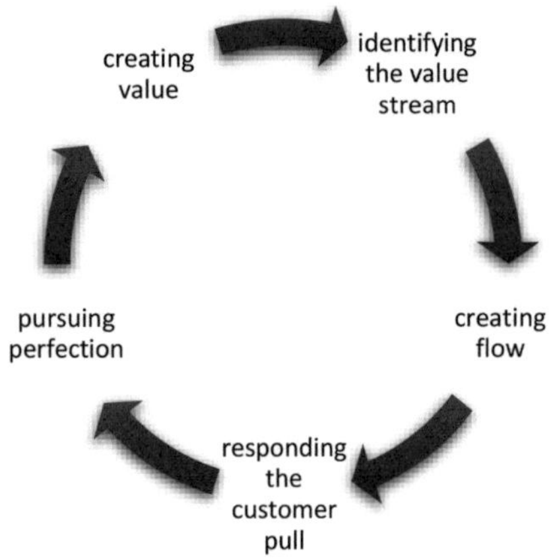

Figure 1.
Five Principles of Lean Manufacturing, adapted from [1].

Waste type	Main effects
Overproduction	More pollution and more raw material use
Waiting	More energy consumption, risk of damage on materials, overuse of space
Transport	More energy consumption, risk of damage on products
Over processing	More material and energy consumption, more pollution
Inventory	More material consumption, risk of damage on products, overuse of space
Movement	RISK of damage on equipment and product, overuse of energy
Defects	More resource consumption, rework, less recycling opportunities

Table 2.
Seven Types of Waste and Their Effects, adapted from [14].

Considering this definition of green manufacturing, lean and green production have common features as both endeavors to diminish waste and to boost the efficiency of production operations [30–34]. Green manufacturing is seen as the positive side-effect [34], a natural extension of lean manufacturing [25] in academic literature.

2.3 Lean manufacturing and its impact on environmental performance

Since its introduction, lean manufacturing has become a significant model in manufacturing industry as a successful process in contributing to competition capacity of organizations [35]. Nevertheless, the growth in ecological concerns due to climate change, environmental degradation and exhaustion of natural resources has compelled the manufacturing establishments to take steps beyond organizational quality by adopting more environmentally sustainable activities and strategies. Therefore, the study of lean and green production has grown into a crucial component of the environment protection agenda [2, 36, 37] and the relationship between these two concepts along with their impacts on environmental performance captured the attention of scholars.

Most studies refer that lean manufacturing supports environmental performance of organizations. Evidence shows that adoption of lean systems minimizes waste and pollution [32, 34, 38]. With its fundamental principle being zero waste [39], lean manufacturing reduces emissions of contaminants by decreasing many types of operational waste such as disposed materials and unnecessary use of energy or water [40]. Furthermore, waste reduction aims of lean are inherently ecological such as unnecessary transportation of products or raw materials [27, 31, 37], whose elimination reduces the unnecessary use of resources along with operational costs [31], and overuse of inventory whose reduction has both financial and environmental benefits [25].

In addition, organizations engaged in lean systems tend to embrace environmental innovations [41]. Lean practices result in less pollution through lower marginal cost of activities that aim for pollution reduction or through lower cost of finding new ways of pollution prevention. Research indicates that capacity of improving operations is enhanced by adoption of lean practices [42], and that higher search costs may prevent managers from discovering new opportunities for environmental protection and investing in them [43].

Being a multifaceted concept, lean approach corporate several practices that intends to achieve the objective of efficiency. Among them JIT practices are used the most and considered one of the most beneficial lean practice with its tools serving to show environmental wastes in production [44, 45]. Also, studies show that VSM generates less emissions to air, less energy use, and more savings [46, 47] and positively affects the amount of solid waste, hazardous substances, water consumption and water pollution [48, 49]. As for kaizen, also known as continuous improvement, it detects and removes hidden wastes and ameliorates the amounts of hazardous chemicals use, water use, savings, water pollution, solid waste [48], material use and emissions to air [48, 50]. TPM affects positively the use of materials as it optimizes the condition of production equipment which in turn supports more efficient use of raw materials with less waste [20, 51] whereas JIT reduces the consumption of material through increasing quality by reducing inventory [15, 21, 52]. In addition, Vais et al. [53] confirmed that adoption of lean practices and tools such as 5S, kaizen and autonomous maintenance optimizes the number of resources used and products manufactured which improve the environmental performance of organizations.

In contrast to positive effects of implementing lean practices on environmental performance, some studies show that lean manufacturing adoption may cause negative effects [25, 32, 54, 55]. For instance, a study demonstrated that firms may incorporate lean systems but air emissions of volatile organic compounds during the production is not reduced [56]. Furthermore, it is presented that quality may be improved by way of consuming more hazardous substances to secure rust-proofing [57] and that more frequent trips for delivery of supplies generates more greenhouse gas emissions [58].

However, overall there is a strong evidence that there is a positive opinion on the effects of lean manufacturing on improved ecological performance particularly in the context of continuous improvement and waste reduction. Most scientific work on the subject shows that lean production reduces overuse of material and energy by continuously advancing operational productivity.

3. Conclusion

Lean manufacturing is accepted as a worthwhile production technique in many industries by the importance of leading higher efficiency and waste (muda) reduction. Alongside the lean concept, environmentally friendly production strategies have also become an issue of interest. Many companies aim for developing products in a way that reduce environmentally detrimental effects of production such as overconsumption of resources and energy, using of hazardous chemicals, and emissions to air, water and land. Lean and Green manufacturing practices both concentrate waste reduction and thus synergic effects on environment. Several studies presented that lean systems inadvertently create ecological gains.

Concerning the air emissions, the findings are controversial suggesting that organizations should critically manage JIT activities as they may cause more air emissions [39, 59]. Although there are a few environmental drawbacks of lean systems, most of the studies demonstrates that there is a positive correlation between adopting lean manufacturing practices and improving environmental conditions.

Author details

Ruhet Genç
Turkish German University, Istanbul, Turkey

*Address all correspondence to: drgench@gmail.com

References

[1] Womack JP, Jones DT. Lean Thinking: Banish Waste and Create Wealth in Your Organisation. 5th ed. New York: Simon and Shuster; 1996. DOI: 10.1038/sj.jors.2600967

[2] Garza-Reyes JA. Lean and green - a systematic review of the state of the art literature. Journal of Cleaner Production. 2015; 102: 18-29. DOI: 10.1016/j.jclepro.2015.04.064

[3] Garza-Reyes JA, Parker HS, Oraifige I, Soriano-Meier H, Harmanto D. An empirical-exploratory study of the status of lean manufacturing in India. International Journal of Business Excellence. 2012; 5(4): 395-412. DOI: 10.1504/IJBEX.2012.047906

[4] Ohno T. Toyota Production System: Beyond Large-Scale Production. 6th ed. Cambridge, Mass: Productivity Press; 1988. DOI:10.4324/9780429273018

[5] Shah R, Ward PT. Lean manufacturing: context, practice bundles, and performance. Journal of Operations Management. 2003; 21: 129-149. DOI: 10.1016/S0272-6963(02)00108-0

[6] Slack N, Brandon-Jones A, Johnston R. Operations Management. 8th ed. Harlow: Pearson; 2016.

[7] Dieste M, Panizzolo R. On the Relationship between Lean Practices and Environmental Performance. In: Proceedings of the IOP Conference Series: Earth and Environmental Science (ICESD '18); 7-9 February 2018; Paris. DOI: 10.1088/1755-1315/151/1/012034

[8] Bai C, Dhavale DG, Sarkis J. Complex Investment Decisions Using Rough set and Fuzzy c-Means: An Example of Investment in Green Supply Chains. European Journal of Operational Research. 2016;248(2): 507-521. DOI: 10.1016/j.ejor.2015.07.059

[9] Zhu Q, Johnson S, Sarkis J. Lean six Sigma and Environmental Sustainability: A Hospital Perspective. Supply Chain Forum: An International Journal. 2018;19(3):1-17. DOI: 10.1080/16258312.2018.1426339

[10] Langenwalter G. Life is our Ultimate Customer: From lean to Sustainability. Target Magazine. 2006;22(1):5-15.

[11] Monden Y. Toyota Production System - an integrated approach to just-in-time, 1st ed. Atlanta: Engineering and Management Press, Institute of Industrial Engineers; 1998.

[12] Krafcik JF. Triumph of the Lean Production System. MIT Sloan Management Review. 1988;30(1).

[13] Holweg M. The genealogy of Lean Production. Journal of Operations Management. 2007;25(2):420-437. DOI: 10.1016/j.jom.2006.04.001.

[14] Moreira F, Alves AC, Sousa RM. Towards Eco-efficient Lean Production Systems. In: Proceedings of the International Conference on Balanced Automation Systems for Future Manufacturing Networks (BASYS); July 2010; Valencia, Spain. pp.100-108, 10.1007/978-3-642-14341-0_12.

[15] Belekoukias I, Garza-Reyes JA, Kumar V. The impact of lean methods and tools on the operational performance of manufacturing organisations. International Journal of Production Research. 2014;52 (18):5346-5366. DOI: 10.1080/00207543.2014.903348.

[16] Rocha-Lona L, Garza-Reyes JA, Kumar, V. Building Quality Management Systems: Selecting the Right Methods and Tools. Productivity Press, CRC Press, Taylor & Francis, Boca Raton, 2013.

[17] Shah R, Ward P. Defining and developing measures of lean production. Journal of Operations Management. 2007;25(4):785-805. DOI: 10.1016/j.jom.2007.01.019.

[18] Andreadis L, Garza-Reyes JA, Kumar V. Towards a conceptual framework for Value Stream Mapping (VSM) implementation: an investigation of managerial factors. International Journal of Production Research. 2017;55(23):7073-7095. DOI: 10.1080/00207543.2017.1347302.

[19] Womack JP. 2006. Value stream mapping. Manufacturing Engineering. 2006;136 (5):145-156.

[20] Konecny PA, Thun JH. Do it separately or simultaneously—an empirical analysis of a conjoint implementation of TQM and TPM on plant performance. International Journal of Production Economics. 2011;133 (2):496-507. DOI: 10.1016/j.ijpe.2010.12.009.

[21] Shingo S. A Study of Toyota Production System: from an Industrial Engineering Viewpoint. 1st ed. New York: Productivity Press, 1989.

[22] Abdulmalek FA, Rajgopal J. Analyzing the benefits of lean manufacturing and value stream mapping via simulation: a process sector case study. International Journal of Production Economics. 2007;107(1):223-236. DOI: 10.1016/j.ijpe.2006.09.009.

[23] Bhuiyan N, Baghel A. An overview of continuous improvement: from the past to the present. Management Decision. 2005;43(5):761-771. DOI: 10.1108/00251740510597761.

[24] Liker JK. The Toyota Way: 14 Management Principles from the World's Greatest Manufacturer. New York: McGraw-Hill ; 2004.

[25] Franchetti M, Bedal K, Ulloa J, Grodek S. Lean and green: Industrial engineering methods are natural stepping stones to green engineering. 2009; 41(9):24-29.

[26] Berkel R, Willems E, Lafleur M. The relationship between cleaner production and industrial ecology. Journal of Industrial Ecology. 1997;1(1):51-66. 10.1162/jiec.1997.1.1.51

[27] Garza-Reyes JA, Winck Jacques G, Lim MK, Kumar, V, Rocha-Lona L. Lean and green – synergies, differences, limitations, and the need for Six Sigma. In: Proceedings of the IFIP International Conference on Advances in Production Management Systems (APMS); 20-24 September 2014; Ajaccio.

[28] Rehman Minhaj AA, Shrivastava RL. Green manufacturing (GM): past, present and future (a state of art review). World Review of Science, Technology and Sustainable Development. (2013); 10(1/2/3):17-55. DOI: 10.1504/WRSTSD.2013.050784.

[29] Abualfaraa W, Salonitis K, Al-Ashaab A, Ala'raj M. Lean-Green Manufacturing Practices and Their Link with Sustainability: A Critical Review. Sustainability. 2020; 12(3):981. DOI: 10.3390/su12030981.

[30] Verrier B, Rose B, Caillaud E. 2016. Lean and Green Strategy: The Lean and Green House and Maturity Deployment Model. Journal of Cleaner Production. 2016;116:150-156. DOI: 10.1016/j.jclepro.2015.12.022.

[31] Carvalho H, Duarte S, Cruz-Machado V. Lean, Agile, Resilient and Green: Divergencies and Synergies. International Journal of Lean Six Sigma 2011;2(2):151-179. DOI: 10.1108/20401461111135037.

[32] Dües CM, Tan KH, Lim M. Green as the new Lean: how to use Lean practices

as a catalyst to greening your supply chain. Journal of Cleaner Production. 2012; 40:93-100. DOI: 10.1016/j.jclepro.2011.12.023.

[33] Hajmohammad S, Vachon S, Klassen RD, Gavronski I. Lean management and supply management: their role in green practices and performance. Journal of Cleaner Production. 2012; 39:312-320. DOI: 10.1016/j.jclepro.2013.06.038.

[34] King AA, Lenox MJ. Lean and Green? An empirical examination of the relationship between lean production and environmental performance. Production and Operations Management. 2001; 10: 244-256. DOI: 10.1111/j.1937-5956.2001.tb00373.x.

[35] Hines P, Holweg M, Rich N. Learning to evolve: a review of contemporary lean thinking. International Journal of Operations and Production Management. 2004; 24(10):994-1011. DOI: 10.1108/01443570410558049.

[36] Cherrafi A, Elfezazi S, Chiarini A, Mokhlis A, Benhida K. The integration of lean manufacturing, Six Sigma and sustainability: a literature review and future research directions for developing a specific model. Journal of Cleaner Production. 2016; 139:828-846. DOI: 10.1016/j.jclepro.2016.08.101.

[37] Garza-Reyes JA. Green lean and the need for six sigma. International Journal of Lean Six Sigma. 2015; 6(3):226-248. DOI: 10.1108/IJLSS-04-2014-0010.

[38] Larson T, Greenwood R. Perfect Complements: Synergies Between Lean Production and eco-Sustainability Initiatives. Environmental Quality Management. 2004; 13(4): 27-36. DOI: 10.1002/tqem.20013.

[39] Rothenberg S, Pil F, Maxwell J. Lean, Green, and the Quest for Superior Environmental Performance. Production and Operations Management. 2001; 10(3):228-243. DOI: 10.1111/j.1937-5956.2001.tb00372.x.

[40] Mackelprang AW, Nair A. Relationship Between Just-in-Time Manufacturing Practices and Performance: A Meta-Analytic Investigation. Journal of Operations Management. 2010; 28 (4):283-302. DOI: 10.1016/j.jom.2009.10.002.

[41] Mollenkopf D, Stolze H, Tate W, Ueltschy M. Green, lean, and global supply chains. International Journal of Physical Distribution & Logistics Management. 2010; 40(1/2):14-41. DOI: 10.1108/09600031011018028.

[42] Womack J, Jones D, Roos D. The Machine that Changed the World. New York: Rawson Associates; 1990.

[43] Jensen R. Adoption and Diffusion of an Innovation of Uncertain Profitability. Journal of Economic Theory. 1982;27(1): 182-193. DOI: 10.1016/0022-0531(82)90021-7.

[44] Aguado S, Alvarez R, Domingo R. 2013. Model of efficient and sustainable improvements in a lean production system through processes of environmental innovation. Journal of Cleaner Production. 2013; 47:141-148. DOI: 0.1016/j.jclepro.2012.11.048.

[45] Florida R. Lean and green: the move to environmentally conscious manufacturing. California Management Review. 1996; 39(1): 80-105. DOI: 10.2307/41165877.

[46] Cherrafi A, Elfezazi S, Govindan K, Garza-Reyes JA, Benhida K, Mokhlis A. A framework for the integration of Green and Lean Six Sigma for superior sustainability performance. International Journal of Production Research. 2017; 55(15):4481-4515. DOI: 10.1080/00207543.2016.1266406.

[47] Deif AM. A system model for green manufacturing. Journal of Cleaner Production. 2011; 19(14),1553-1559. DOI: 10.1016/j.jclepro.2011.05.022.

[48] Pampanelli AB, Found P, Bernardes AM. A lean & green model for a production cell. Journal of Cleaner Production. 2014; 85:19-30. DOI: 10.1016/j.jclepro.2013.06.014.

[49] Fliedner G. Sustainability: a new lean principle. In: Proceedings of the 39th Annual Meeting of the Decision Sciences Institute. Baltimore, Maryland; 2008. p. 3321-3326.

[50] Garza-Reyes JA, Kumar V, Chaikittisilp S, Tan KH. The effect of lean methods and tools on the environmental performance of manufacturing organisations. International Journal of Production Economics. 2018; 200:170-180. DOI: 10.1016/j.ijpe.2018.03.030.

[51] Eti MC, Ogaji SOT, Probert SD. Implementing total productive maintenance in Nigerian manufacturing industries. Applied Energy. 2004; 79(4): 385-401. DOI: 10.1016/j.apenergy.2004.01.007.

[52] Cua KO, McKone-Sweet KE, Schroeder RG. Improving performance through an integrated manufacturing program. Quality Management Journal. 2006; 13(3):45-60. DOI: 10.1080/10686967.2006.11918561.

[53] Vais A, Miron V, Pederson M, Folke J. Lean and green at a Romanian secondary tissue paper and board mill-putting theory into practice. Resources, Conservation and Recycling. 2006; 46(1): 44-74. DOI: 10.1016/j.resconrec.2005.06.005.

[54] Sanchez Rodrigues V, Kumar M. Synergies and Misalignments in Lean and Green Practices: A Logistics Industry Perspective. Production Planning & Control. 2019; 30(5-6): 369-384. DOI: 10.1080/09537287.2018.1501812.

[55] Sartal A, Martinez-Senra AI, Cruz-Machado V. Are All Lean Principles Equally Eco-Friendly? A Panel Data Study. Journal of Cleaner Production. 2018; 177: 362-370. DOI: 10.1016/j.jclepro.2017.12.190.

[56] Venkat K, Wakeland W. Is lean necessarily green?. In: Proceedings of the 50th Annual Meeting of the ISSS; June 2006; Sonoma, CA, USA. Available from: https://journals.isss.org/index.php/proceedings50th/article/view/284

[57] Helper S, Rozwadowski H, Clifford PG. Can Green Be Lean?. In: Proceedings of Academy of Management Annual Meeting, Organizations and the Natural Environment; 10-13 August 1997; Boston.

[58] Katayama H, Bennet D. Lean production in a changing competitive world: a Japanese perspective. International Journal of Operations & Production Management. 1996; 16(2):8-23. DOI: 10.1108/01443579610109811.

[59] Simons D, Mason R. Lean and green: 'doing more with less'. International Commerce Review: ECR Journal. 2003;3(1):84.

Chapter 14

Circular and Lean Food Supply Chains

Stella Despoudi

Abstract

Circular economy (CE) refers to the industrial economy that aims to achieve enriched sustainability through restorative objects and supply chain design. Many governments have put in place different initiatives in line with the CE. On the other hand, the term Lean operations refers to the reduction of the non-value adding activities and waste in a supply chain. The food sector has been criticized for its sustainability and circularity due to the high levels of food and packaging waste and at the same time the increasing costs. Although food supply chain entities have started to implement circular economy and lean practices, the current efforts do not seem to be sufficient to achieve a circular and lean food system. The aim of this chapter is to explore the possibility of a circular and at the same lean food supply chain.

Keywords: circular economy, lean supply chain, food supply chain, sustainability

1. Introduction

In the past 20 years, interest in the notion of supply chain management has been growing rapidly in both global academia and business world. Behind this are the rapid changes in the global economic environment, the increasing variety of products, the increasing demand for delivery deadlines, the shorter product life cycle and the increasing expectations of consumers for products and services [1]. At the same time, with the continuous development of e-commerce, globalization and offshoring of production, the complexity and related risks of supply chain management also increase [2]. These new factors also promote the increasing interest of enterprises in supply chain management and re-examine their supply chain strategy.

In the early view of supply chain management, the scope of supply chain management is only discussed within the enterprise, and then it gradually derives to the relationship and cooperation between the enterprise and the upstream and downstream enterprises [3]. Moreover, it is no longer confined to the manufacturing industry, for a variety of industries, even the service industry, supply chain management has shown its significance. Today, supply chain management has become a more comprehensive concept. It covers all the processes from raw material suppliers to end customers. All the links that have an impact on cost and satisfying end customer needs are included in the scope of discussion by supply chain management.

However, at the same time, due to the increasing global environmental issues, it also puts forward new topics and higher requirements for modern supply chain management [4]. Since the industrial revolution in the 18th century, human society has undergone tremendous changes with the development of science and technology. In this process, the production technology is progressing rapidly, the scale of production is expanding constantly, and the production efficiency is advancing rapidly [5–7]. However, the progress of society and science and technology not only brings great economic benefits to people, but also has a tremendous impact on the environment. People blindly pursue economic benefits, thus excessive exploitation of natural resources, not only lead to regional environmental pollution, but also make some global environmental problems such as global warming, biodiversity reduction and marine pollution [8, 9]. Moreover, statistics show that more than 2.4 billion people have been affected by climate-related disasters from 2002 to 2012, an increase of 700 million over the last decade [10]. These environmental issues have threatened the living environment of human beings and will affect the sustainable development of society. As a result, people began to pay attention to the seriousness of environmental problems and try to seek measures for improvement and sustainable development.

Governments in various countries have promulgated laws and regulations to regulate green manufacturing in enterprises in the early 21st century. The EU has promulgated RoHS, WEEE and REACH directives, which regulate the supply chains of different industries from different aspects, requiring them to become a greener supply chain [11]. Meanwhile, some international NGOs are also involved in this series of actions. The International Organization for Standardization has developed ISO14000 series environmental management standards to manage and plan the activities, products and services of all organizations, including global enterprises and social groups [12].

Academia has also paid more attention to and discussed the environmental responsibility of enterprises. Although there are many debates on corporate social responsibility, most scholars agree that enterprises, as an important factor of social composition, are obliged to undertake certain environmental responsibilities [13]. Elkington [14] also put forward the theory of triple bottom line in 1998, which expresses that corporate profit, social responsibility and environmental responsibility are the foundation of enterprises, and also the foundation of enterprises' continuous development and long-term development [14].

Finally, with the waste of resources and the increasingly serious environmental pollution, consumers have also begun to pay more attention to the environmental protection behind products and services [15, 16]. The change of environmental willingness of American consumers from 2008 to 2013. It can be found that 71% of consumers will take environmental factors into account when shopping in 2013, up from 66% in 2008. This means that companies need to meet the growing environmental needs of consumers, so companies also need to make the supply chain greener, which can also enable them to have more competitive advantages in the market competition. In the context of these three factors, enterprises need to make their supply chains more environmentally friendly and sustainable, whether from the perspective of government mandatory supervision, or to meet the needs of consumers, or for the sustainable development of companies.

As an industry closely related to people's daily lives, food industry has huge direct or indirect impact on the environment. Moreover, as an important part of the global economy, the food supply chain is also closely related to the consumption of natural resources and the emission of pollution [17]. At the same time, the research also shows that activities related to food packaging, transportation and waste disposal account for 5–10% of global greenhouse gas emissions, and this figure is expected to continue to rise [18]. On the other hand, food safety is an important topic of global concern, and also the top priority of the food industry. This means that for food companies, under the important premise of ensuring food safety, they also need to solve the challenges from the environment, so that their supply chain need to have as little impact on the environment as possible [19]. As a result, an environmentally friendly and sustainable supply chain is particularly important for the food industry.

In such an overall environment, many new concepts about supply chain management have emerged, including green supply chain management, sustainable supply chain management, closed-loop supply chain, lean supply chain, reverse logistics and so on. These concepts provide enterprises with a more environmentally friendly supply chain development model from different perspectives, enabling them to re-examine their supply chain strategy [20]. Similarly, circular economy, as a relatively new concept, emphasizes the sustainable use of resources and energy in the economic activities, thereby minimizing its impact on the environment [21]. Another supply chain practice which is hugely involved in the manufacturing industry is the philosophy of Lean production systems from Toyota [22]. The purpose of lean supply chain management to eliminate waste and non-value adding activities by optimizing the supply chain by adopting a wide range of lean practices/tools (e.g. Total productive maintenance, just in time, 5 s, Total quality maintenance and Kanban card). Both lean and circular economy are becoming essential elements for successfully managing supply chains and in particular food supply chains. However, their interrelationship has not been widely discussed especially in the food sector.

2. Lean and circular supply chains

This section starts with a definition of lean production and a discussion of the origins and practices of lean. This is followed by a discussion of the lean tools. After that, the definition of circular economy, the practices of circular economy i.e. 3 Rs, and the definition of the circular supply chain. Then the characteristics of the food supply chain are discussed. The chapter concludes with a critical discussion of the circular and lean supply chain characteristics.

2.1 Definition of lean

Krafcik [23] first introduced the lean concept after studying global automotive manufacturing, with a specific focus on deciphering the inner workings of Toyota's production system. At their core, lean practices aim to streamline the flow of value by systematically reducing waste during the manufacture of a product [23]. However, it was Womack et al. [24] who first produced a consolidated study that claimed that five core lean principles could be applied to every industry [24]. They are:

1. Specifying value creation

2. Identifying the value streams of the production process and eliminating waste

3. Creating flow in the production line from supplier to customer

4. Creating pull, by allowing customer demand to be the driver

5. Striving to achieve the four aforementioned principles through a systematic approach towards continuous improvement

Whilst the lean concept is nothing new, it was observed by Baker [25] that there are few organizations which fully understand the underlying philosophy behind its key principles [25]. However, Liker [26] identified that for lean production practices to be effective, they needed to be rooted in a people-centric system where employees are directly involved in the identification and implementation of continuous improvement [26]. This organizational approach was also identified by Hines et al. [27], who focused on the idea of Total Quality Management (TQM) [27]. TQM requires the development of an overall organizational culture which moves beyond the production floor. This is demonstrated by the myriad of businesses who have unsuccessfully attempted to implement lean practices. This organisational approach was also identified by Hines et al. [27], who focused on the idea of Total Quality Management (TQM) [27, 28]. Academic studies have looked at such failures, identifying some common factors as responsible for businesses being unable to successfully enact them. These are: the complexity of lean management implementation; the limitations on a successful impact caused as a result of contingency factors; a focus on Just-In-Time (JIT) practices without sufficient consideration for other key aspects of Operations Management (OM); and the lack of attention paid to Human Resource Management (HRM) Further, the difference between those that were successful and unsuccessful was the implementation of 'soft' lean practices, such as small group problem solving, training to complete multiple tasks, supplier partnerships, customer involvement, and continuous improvement.

2.2 Origin of lean production and lean principles

In retrospect, the term Lean Production System (LPS) was introduced into world industry in order to enhance and maximize efficiency throughout manufacturing processes which became a suitable model for modern manufacturing companies to design processes and procedures in their firms. According to Western industry and Japanese industry, Western world industry employed automation, production systems and computer-aided technologies to enhance manufacturing process whereas the Japanese industries invented a customer-value focused method called "Lean Manufacturing" which also known as "Lean production" [29]. It has become a major driving force of modern business' success and has widely accepted by the companies [30].

The philosophy of Lean was originated by Toyota in the 1950s by Taiichi Ohno. The techniques aim to eliminate all the wastes and excess (called as Muda) from the production system. Lean Production has principles to avoid eight causes of waste which are storage, transport, accessibility of processes, unnecessary movement, waiting times, overproduction, tight tolerances, defects and lastly, unused skills of the employees. Lean principles were further developed which includes the following techniques: value of customers, value stream mapping, flow production capability, pull system and eliminate all forms of waste. More supportive evidence by Bicheno and Holweg [31] state wastes as "non-value added" to the companies which are the factors needed to be minimized to zero as much as possible and maintain value-added activities (what customers are willing to pay for) in the same or higher level [31]; Practically, it's inevitable in any businesses that wastes will occur.

VDI [32] commented on lean production as "an enterprise-specific methodical system of rules for the continues orientation of all enterprise processes to the customer in order to achieve the largest by the enterprise management" which can lead to a continuous improvement process (CIP) internally [32]. CIP was defined as a "frame of thought and the proceeding actions for formalized, targeted and progressive improvement of activities." and became significantly important to all firms who interested in leans [33]. In order to be successful in terms of continuous improvement of production, integration of the following principles below is needed where the Lean Production principles' symbols shown by House of Lean Production

in and Tripathi, D. [34], "Relationship between TQM and TPM implementation factors and business performance of manufacturing industry in Indian context", International Journal of Quality & Reliability Management [35]. However, all of these cannot be done or implemented if a company has a wrong mindset with business culture, leadership and coaching.

2.3 Lean production tools/practices

This section will discuss six lean production tools/practices which are: 5 s Total productive maintenance (TPM), Total quality management (TQM), Value stream mapping (VSM), Kanban, and Just in time (JIT).

One of the important topics in implementing lean is lean production tools since lean principles were mentioned in the previous part, to find appropriate tools to be used in organization is essential to achieve those principles. An interesting quote from Abraham Maslow in 1966 for his hierarchy of motivation "It is tempting, if the only tool you have is a hammer, to treat everything as if it were a nail" so as lean tools which all firms need to identify their objective before using any lean tools. This can be observed on SMEs companies; these firms will firstly focus on the most necessary lean practices rather than a piecemeal one where most of them start implementing lean by using these tools such as 5 s, Kanban and Total productive maintenance (TPM). "SMEs should start with minimal financial investment on lean practices such as, Kanban, 5s and Honshin" due to their incapability to implement everything at once, so it's exceptional to choose the easiest or cheapest. 5S, one of the most common tools in lean because it's easy to do without the needs of expense. As its name, it has 5 of S which are, Sort, Set-in-order, Scan, Standardize and Sustain. The objective to reorganize workplaces and create a strong mindset for an entire company as a whole in order to maintain workplaces in good condition and promised the employees to be more self-discipline [35]. Another related practice to 5 s is called Total productive maintenance (TPM), "a maintenance program which involves a newly defined concept for maintain plants and equipment" [36].

The benefits of using TPM is to minimize defects, breakdown and accidents as much as possible in all functional workplaces, also involve entire organization to work together as a team by sharing knowledge and experience. Moreover, it shares the same principle to 5 s, which is to keep the workplace clean and sustainable [36]. One of the tools that are worth to mention is Total quality management (TQM), the tool that shares similarities with TPM and often used interchangeably [37]. The main concept of TQM is to have continuous improvement in all work; it means that from strategic planning and decision-making to the execution in the shop floor [38]. The principle of TQM is that mistaken from human, which might be occurred from faulty systems, needed to be prevented from a repetition of the same mistake by establishing the new process.

The next tool is called Value stream mapping (VSM), an effective tool for the practice of lean manufacturing. It is used to visualize the steps needed from the beginning of product creation until delivering to end-customer. The map will help us to identify where in the process is a bottleneck so operators could identify and improve it. Lastly, a well-known lean method called Kanban, aims to minimize inventory at any one time [39]. It is used to indicate when the demand of product is available in the specific circumstances so production can operate in the right time. Kanban system is very popular in Japanese companies due to the cost saving by avoiding overproduction, improving flexibility in each work stations, reducing inventory stock, waiting times and logistics costs. Kanban is one of the methods used to achieve Just-in-time manufacturing (JIT), a management philosophy implemented and designed by Taiichi Ohno, 1998. Just-in-time (JIT) is most widely adopted and practices recently in several organizations, which aims to have the right items with certain quantity and quantity in the right place and time [40].

2.4 Definition of circular economy

The early thought of circular economy can be tracked back to 'the spaceship theory' mentioned by Boulding in 'Economics of the coming spaceship earth' in 1966. Boulding [41] stated that in an isolated and closed spacecraft, due to limited resources, if resources are not recycled, it will eventually go to destruction, the same is true of the earth [41]. After Boulding put forward this theory, people began to realize the importance of sustainable development to human beings. Then, in 1989, Pearce and Turner formally and systematically put forward the concept of circular economy, and explained the transition from traditional linear economy to circular economy.

However, since the different proposed antecedents and scopes, different literature give different answers to the definition of circular economy. Circular economy is an economic development mode that minimizes material, energy and environmental damage without restricting economic growth and social and technological progress. Hobson [42] defined circular economy as a regenerative industrial system, which aims to eliminate waste and be regenerate mainly through better design of material, system and business model [42]. Murray et al. [43] claimed that circular economy is an economic model that achieves ecosystem optimization through rational design and management of resources, procurement, production and recycling processes [43].

Although there is no comprehensive and commonly accepted definition of circular economy at present, some common goals are revealed in relevant literature. First of all, most of the definitions of circular economy in the literature mention that circular economy is used to replace the traditional linear economy, and through intentional design and management to make the economic model restorative [44]. Govindan and Hasanagic [45] pointed out that because the traditional linear economy did not consider the impact of natural resources and human resources, such as social capital, the use of circular economy to replace the linear economy can make resources reflect not only economic value but also environmental and social value [45]. Secondly, relevant literature also point out that circular economy is to minimize the waste and damage of natural resources and environment caused by economic development. Thirdly, circular economy aims to improve economic, social and environmental values rather than pursuing economic values alone [46]. Last but not least, at the macro level, circular economy system are more resilient because they rely less on external resources [47]. A common point of most circular economy definitions: closed flow of materials and the use of materials in multiple stages. As a result, the above five points explain the goal of circular economy, and this also gives a clearer framework and direction for this research's further discussion. Although there is no comprehensive definition of circular economy at present, considering the characteristics of supply chain, this study will follow Murray et al.'s definition: circular economy is an economic model that achieves ecosystem optimization through rational design and management of resources, procurement, production and recycling processes.

2.4.1 "3R" principles

The 3R principles refer to reduction, reuse and recycle. As the core principles in the practice of circular economy, 3R principles have been widely discussed in many studies [48]. In the Circular Economy Promotion Law promulgated by the

Chinese government in 2008, the Chinese government also directly defined circular economy as a reduction, reuse and recycling activity in the process of production, circulation and consumption [49]. Thus, it can be seen the 3R principles are of great significance in the practice of circular economy.

- Reduction principle

  Reduction is the first principle in the 3R principles, which means reducing resource consumption and waste generation through better design or management at the production and consumption stage [50]. The purpose of reduction principle is to improve from the source of production stage, so as to reduce materials and energy entering the production and consumption stage. Therefore, the reduction principle is aimed at requiring businesses to start considering material and energy savings and waste reduction at the input end, rather than making great efforts on recycling or waste disposal stage after the waste is generated. For enterprises, there are many measures to implement reduction principle. Using environmentally friendly raw materials, environmentally friendly production methods, simpler packaging design and more compact and lightweight product design are all effective ways for enterprises to implement reduction principle. Enterprises can also better apply the principle of reduction by improving their ecological efficiency, although ecological efficiency also plays a positive role in the other two principles.

- Reuse principle

  Reuse principle refers to the reuse of product or product packaging in an initial form as many times as possible and in various forms, so as to prevent it from becoming waste too early. Extending the service life of products as far as possible is also a requirement of the reuse principle. The implementation of the reuse principle means that companies can save a lot of resources, labor and energy to manufacture new products or components and packaging of products [51]. However, the principle of reuse also puts forward a higher requirement for companies. Prendeville et al. [52] pointed out that the company not only needs to be able to design products with longer service life, but also needs to consider how to encourage consumers to recycle and reuse products or packages [52]. Therefore, it also shows that the principle of reuse not only puts forward certain requirements for enterprises, but also depends on the attitude and participation of consumers. In the process of implementing reuse principle, all stakeholders need to participate, including consumers' recycling and reuse of products and packaging.

- Recycle principle

  The recycle principle refers to the recycling of wastes as much as possible through the reprocessing of wastes, so that they can be converted into useful resources and thus reduce the generation of non-recyclable garbage [53]. There are also two ways of recycling wastes: primary recycling and secondary recycling. Henshaw et al. [54] mentioned that primary recycling refers to recycling and processing waste to form new products or packages identical to the original (such as using scrap steel to produce new steel) [54], while secondary recycling refers to using waste as raw materials for other products with different properties without considering the original properties of waste. The principle of recycling can be used to control the end of the waste generation, so that the waste can be re-entered into the production or consumption stage, which also reduces the possible environmental pollution. However, once a company pays too much attention to the recycling process and achieves good results, it will not be interested in controlling the amount of waste.

2.5 Definition of circular supply chain

Under the trend of globalization, the importance of supply chain management has become increasingly prominent. Many scholars even emphasize that the competition among enterprises today is the competition among supply chains [55]. At the same time, how to make the supply chain greener and more sustainable has become a key issue under the situation that all parties are paying more and more attention to environmental protection [56, 57]. With the continuous development of circular economy concept, an increasing number of scholars begin to study the application and practice of circular economy in various fields and levels. As a result, the application of circular economy in supply chain has naturally become a hot research topic.

When discussing the concept of circular supply chain, many scholars will compare several concepts of supply chain sustainability with that of circular supply chain [58]. These concepts related to supply chain sustainability are mainly reverse logistics, green supply chain, sustainable supply chain and closed-loop supply chain. Although these concepts cannot fully contain the meaning of the circular supply chain, the four concepts each contain some scopes and circular flows of the circular supply chain. Reverse logistics enables products or materials to flow reversely from consumers, the end point of the supply chain, to each node of the supply chain along the supply chain channel, so that discarded or damaged products can regain their value and avoid becoming waste. The definition of green supply chain refers to the purpose of minimizing the impact on the environment from raw material acquisition, processing, production, packaging, warehousing, transportation, consume to scrap disposal [59]. Sustainable supply chain management emphasizes a broader and comprehensive supply chain strategic management, taking into account the environmental, economic and social factors in supply chain management, so as to achieve the goal of long-term sustainability of the supply chain [60]. Finally, as a relatively new concept, closed-loop supply chain integrates the forward and reverse supply chains and covers the whole life cycle of products from cradle to grave, in order to close the flow of materials and reduce pollution and waste generation [61].

The above concepts of supply chain sustainability include some circular flows of the circular supply chain, but not all scopes and circular flows. The model of circular supply chain needs to be extended on the basis of closed-loop supply chain, and the extension point is the scope and focus of material recovery system. For the scope, post-production management should be taken into account in the circular supply chain so as to include open-loop in the supply chain; for the focus, the value chain system of the circular supply chain also needs to be derived from the related by-product synergies and waste recovery flows. As a result, Batista et al. [58] defined the circular supply chain as "The coordinated forward and reverse supply chains via purposeful business ecosystem integration for value creation from products or services, by-products and useful waste flows through prolonged life cycles that improve the economic, social and environmental sustainability of organisations."

2.6 Characteristics of food supply chain

Food industry as an industry closely related to people's daily life, because of the particularity of food itself, food supply chain also has many different characteristics. Maloni and Brown [62] pointed out that due to the food safety and hygiene, the product quality requirements of food supply chain are strictly regulated [62]. At the same time, food supply chain is also highly dependent on the environment because of its vulnerability to deterioration. Similarly, due to the limitation of shelf life, the change of raw materials, semi-finished products and final product quality change is also a challenge for the food supply chain. Moreover, Papaioannou et al. [63] also stated that the food supply chain has inherent uncertainties in demand, process and suppliers [63]. First of all, it is not easy to forecast the demand of customers, and the food supply chain also needs to face further challenges from the disturbance of weather changes and customer preference changes. At the same time, due to yield changes, seasonal factors, some perishable food and other factors, its supply and production and storage process will also face certain uncertainties and risks.

3. Circular and lean food supply chain

Today's business environment is a very dynamic one with the rapidly changing consumer preferences, which pressures the business to respond to customers' needs and wants, become more sustainable, and cost efficient. In recent years, sustainability has become an emerging goal of food supply chain due to the sustained attention of various aspects to environmental issues. According to Consumers are increasingly interested in the sustainability behind products (raw materials, procurement, packaging and transportation). In addition, due to the pressure of government regulation and consumers' increasing awareness of environmental issues in consumption, many enterprises are forced to turn their supply chain model into a circular supply chain. However, the circular food supply chain can also bring many other benefits to the company. For instance, Beske et al. [64] pointed out that circular supply chain can help companies provide consumers with higher quality and safer food [64]. Geissdoerfer et al. [65] also mentioned that the circular supply chain can reduce food waste, improve brand benefit and increase long-term profits [65–67]. As a result, food companies need to use a circular supply chain to improve their customer loyalty and competitive advantage.

At the same time, some scholars also discussed challenges faced by the circular food supply chain. Carbon emissions and other environmental pollution of the food supply chain (waste cooking oil supply chain) after the implementation of the circular supply chain and concluded that the impact of the supply chain on the environment has been significantly reduced, but the short-term economic benefit is still a challenge. Food waste is caused by food packaging, so food packaging is a challenge for the circular supply chain. In addition, many farmers in developing countries generally lack sustainable knowledge, which may affect the implementation of the circular supply chain Finally, weak regulation of laws and regulations, lack of pressure from the market and lack of support from other stakeholders in the supply chain are all challenges faced by the circular food supply chain. Circular economy brings the advantages renewability, reusage and recycling that the food sector needs.

On the other hand, lean supply chain provides food companies with a series of practices that optimize supply chains through more efficient routes and usage of materials, cost efficiency and better management of customer demand. Due to the simplicity of lean principles, high productivity, short lead-time, and improved overall efficiency were able to replace the decline of Computer Integrated Manufacturing (CIM) and became the standard of today's manufacturing practice.

Many research studies carried out in manufacturing industries showed that the owners are likely to adopt lean principles to the production due to the advantages that they acquire regardless of its scale. This is because either small or large business has its advantages of lean principles. The biggest size of SMEs is taken place in China with less than 999 employees, where in EU & UK estimated should be less than 250 employees. SMEs aim to have a better relationship with higher flexibility and strong work communication internally to have a simple and clear infrastructure for everyone to follow. On the contrary, large businesses aim to find access to resources and exploit them which will allow them to capitalize a market with a better lean supply chain than SMEs.

Lean production can help to reduce inventories for manufacturers since high inventories can result in inventory costs and available workplaces are minimized. Also, it can decrease any unnecessary times in supply chains such as lead times for customers, changeover time in using machineries and idle time of workers and educate managerial skills to upper managers and improved unskilled workers in the production. Applying lean principles, company will gain advantages in terms of increasing availability of machines and reducing manpower, bettering work environment in terms of higher safety and more organized workplace and also, more cost saving and higher profits as all potential wastes are eliminated, all hidden will also be eliminated.

Nevertheless, even though there are many advantages from applying lean principles, there are barriers obstructing lean productions to be implemented, which taken in accounts every organizations. Lean production is not suitable for applications that have high variations of customer demands and highly customized product. This is because lean principle aims to balance the flow production, which cannot respond to the rapid change in demand fluctuation. Thus, the lean idea such as Just-In-Time (JIT) can be ambiguous to be implemented in the production. Availability of supplies can also cause trouble since lean aims to keep only a small amount of inventory in the production in order to minimize unnecessary cost. This will trouble companies if unexpected circumstances occur such as employee strikes, error on products and transportation delays which can damage customer relationships and whole supply chains. Another factor is financial plan since it needs huge investment to develop regard to training cost, purchasing cost and takes huge amount of time to revolutionize entire organizations to have strong mindset; some employees are not willing to change their traditional way of working and afraid to make any mistakes regard to the new system.

4. Conclusion

The purpose of this chapter was to discuss the possibility of having circular and lean food supply chains. The chapter started with an introduction to the need for circular and lean food supply chains. Then this was followed with the definition of lean production, a discussion of the origins and practices of lean and the lean tools. After that, the definition of circular economy, the practices of circular economy i.e. 3 Rs, and the definition of the circular supply chain were explained. Then the characteristics of the food supply chain were discussed. The chapter concludes with a critical discussion of the circular and lean supply chain characteristics. It is evident that both approaches can be combined in order to bring the ideal and desirable food supply chain of the future. Of course, none of them is panacea, and therefore food companies that wish to implement such practices need to adapt them to their own needs and abilities.

Author details

Stella Despoudi[1,2]

1 University of Western Macedonia, Grevena, Greece

2 Aston University, Birmingham, UK

*Address all correspondence to: aff00841@uowm.gr

References

[1] Stadtler, H. (2014). Supply Chain Management: An Overview. Springer Texts in Business and Economics, pp.3-28.

[2] Gereffi, G. and Lee, J. (2012). Why the World Suddenly Cares About Global Supply Chains. Journal of Supply Chain Management, 48(3), pp.24-32.

[3] Simatupang, T. and Sridharan, R. (2005). The collaboration index: a measure for supply chain collaboration. International Journal of Physical Distribution & Logistics Management, 35(1), pp.44-62.

[4] Fahimnia, B., Sarkis, J. and Davarzani, H. (2015). Green supply chain management: A review and bibliometric analysis. International Journal of Production Economics, 162, pp.101-114.

[5] Spanaki, D, Karafili, E, Despoudi, S. (2019) Data Sharing in Agriculture 4.0: Concepts and Challenges, EurOMA Conference 2019 – Finland.

[6] Spanaki, K; Sivarajah, U; Fakhimi, M; Despoud, S; Irani, Z, (2021a) Disruptive Technologies in Agricultural Operations: a meta-review of AgriTech and AI Research, Annals of Operations Research, In press.

[7] Spanaki, K; Karafili, E; Sivarajah, U; Despoud, S; Irani, Z (2021b) Artificial intelligence and food security: swarm intelligence of AgriTech drones for smart AgriFood operations, Production Planning and Control, In press.

[8] Despoudi, S, Papaioannou, G, Dani, S (2020a) Producers Responding to Environmental Turbulence in the Greek Agricultural Supply Chain: Does Buyer Type Matter?, Production Planning and Control, https://doi.org/10.1080/09537287.2020.1796138

[9] Despoudi, S., (2020b). Green Food Supply Chain, In Galanakis, C. (Ed) Food Industry and the Environment https://doi.org/10.1016/B978-0-12-816449-5.00002-3.

[10] Munich Re (2013). RISKS AND CHANCES OF CLIMATE CHANGE. [online] Available at: https://iumi.com/images/Berlin2015/3Press ies/16.09._09.40_Hoeppe_Handout.pdf

[11] Koh, S., Gunasekaran, A. and Tseng, C. (2012). Cross-tier ripple and indirect effects of directives WEEE and RoHS on greening a supply chain. International Journal of Production Economics, 140(1), pp.305-317.

[12] Montabon, F., Melnyk, S., Sroufe, R. and Calantone, R. (2000). ISO 14000: Assessing Its Perceived Impact on Corporate Performance. The Journal of Supply Chain Management, 36(2), pp.4-16.

[13] Bondy, K., Moon, J. and Matten, D. (2012). An Institution of Corporate Social Responsibility (CSR) in Multi-National Corporations (MNCs): Form and Implications. Journal of Business Ethics, 111(2), pp.281-299.

[14] Elkington, J. (1998). ACCOUNTING FOR THE TRIPLE BOTTOM LINE. Measuring Business Excellence, 2(3), pp.18-22.

[15] Despoudi, S., Papaioannou, G., Saridakis, G. & Dani, S. 2018. Does collaboration pay in agricultural supply chain? An empirical approach. International Journal of Production Research, 56, 4396-4417.

[16] Despoudi, S., 2016. An Investigation of the Collaboration – Postharvest Food Loss Relationship and the Effect of the Environmental Turbulence Factors. Loughborough University. https://

dspace.lboro.ac.uk/dspace-jspui/bitstream/2134/21785/1/Thesis-2016-Despoudi.pdf

[17] Halkier, B. (2001). Risk and food: environmental concerns and consumer practices. International Journal of Food Science and Technology, 36(8), pp.801-812.

[18] Vermeulen, S., Campbell, B. and Ingram, J. (2012). Climate Change and Food Systems. Annual Review of Environment and Resources, 37(1), pp.195-222.

[19] Ala-Harja, H. and Helo, P. (2014). Green supply chain decisions – Case-based performance analysis from the food industry. Transportation Research Part E: Logistics and Transportation Review, 69, pp.97-107.

[20] Masi, D., Day, S. and Godsell, J. (2017). Supply Chain Configurations in the Circular Economy: A Systematic Literature Review. Sustainability, 9(9), p.1602.

[21] Ghisellini, P., Cialani, C. and Ulgiati, S. (2016). A review on circular economy: the expected transition to a balanced interplay of environmental and economic systems. Journal of Cleaner Production, 114, pp.11-32.

[22] Vinodh, S., Prakash, N.H. and Selvan, K.E. (2011), "Evaluation of leanness using fuzzy association rules mining", International Journal of Advanced Manufacturing Technology, Vol. 57 No. 1, pp. 343-352.

[23] Krafcik, J., 1988. Triumph of the lean production system. MIT Sloan Management Review, 30(1), pp. 41-52.

[24] Womack, J., Jones, D. & Roos, D., 1990. The Machine That Changed the World. New York: Rawson Associates.

[25] Baker, P., 2002. Why is lean so far off?. Works Management, Volume October, pp. 1-4.

[26] Liker, J., 2004. The Toyota Way: 14 management principles from the world's greatest manufacturer. New York: McGraw Hill Professional.

[27] Hines, P., Martins, A. & Beale, J., 2004. Testing the boundaries of lean thinking: Observations from the legal public sector. Public money and management, 28(1), pp. 35-40.

[28] Bateman, N., 2005. Sustainability: the elusive element of process improvement. International Journal of Operations & Production Management, 25(3), pp. 261-276.

[29] Sanders, A., Elangeswaran, C. & Wulfsberg, J., 2016. Industry 4.0 Implies Lean Manufacturing: Research Activities in Industry 4.0 Function as Enablers for Lean Manufacturing. Journal of Industrial Engineering and Management, 9(3), pp. 811-833.

[30] Kolberg, D. & Zuhlke, D., 2015. Lean Automation enabled by Industry 4.0 Technologies. IFAC Conference, pp. 1870-1875.

[31] Bicheno, J. and Holweg, M. (2009). The Lean Toolbox: The Essential Guide to Lean Transformation. Buckingham: Production and Inventory Control, Systems and Industrial Engineering (PICSIE) Books.

[32] The Association of German Engineers (VDI). (2012). Guideline 2870-1 – Lean Production Systems – Basic principles, introduction and review. Beuth, Berlin.

[33] Thomas MASCHEK, Khashayar KHAZRAEI, Sabine HEMPEN and Jochen DEUSE. (2011). Managing Coutinuous Improvement Process as an Organizational Task in

Course of a Century, Chair of Industrial Engineering, Dortmund, Germany: Technische Universität Dortmund.

[34] Seth, D. and Tripathi, D. (2005), "Relationship between TQM and TPM implementation factors and business performance of manufacturing industry in Indian context", International Journal of Quality & Reliability Management, Vol. 22 No. 3, pp. 256-277.

[35] Sui Pheng, L. (2001). Towards TQM – integrating Japanese 5-S principles with ISO 9001:2000 requirements. The TQM Magazine, 13(5), pp.334-341.

[36] Venkatesh, J. (2005). An Introduction to Total Productive Maintenance (TPM), The Plant Maintenance Resource Center, pp 2-3

[37] Seth, D. and Tripathi, D. (2005), "Relationship between TQM and TPM implementation factors and business performance of manufacturing industry in Indian context", International Journal of Quality & Reliability Management, Vol. 22 No. 3, pp. 256-277.

[38] Hashimi, K. (2012). Introduction and Implementation of Total Quality Management (TQM) - iSixSigma. [online] iSixSigma. Available at: https://www.isixsigma.com/methodology/total-quality-management-tqm/introduction-and-implementation-total-quality-management-tqm/

[39] Rahman, N., Sharif, S. and Esa, M. (2013). Lean Manufacturing Case Study with Kanban System Implementation. Procedia Economics and Finance, 7, pp.174-180.

[40] D. K. Singh and Dr. Satyendra Singh. (2013). JIT: A Strategic Tool of Inventory Management, International Journal of Engineering Research and Applications (IJERA) ISSN: 2248-9622, Vol. 3, Issue 2, March -April 2013, pp.133-136

[41] Boulding, K.E., 1966. The economics of the coming spaceship earth. In: Jarrett, H. (Ed.), Environmental Quality in a Growing Economy: Essays From the Sixth RFF Forum. Routledge, pp. 3-15.

[42] Hobson, K. (2015). Closing the loop or squaring the circle? Locating generative spaces for the circular economy. Progress in Human Geography, 40(1), pp.88-104.

[43] Murray, A., Skene, K. and Haynes, K. (2015). The Circular Economy: An Interdisciplinary Exploration of the Concept and Application in a Global Context. Journal of Business Ethics, 140(3), pp.369-380.

[44] Mendoza, J., Sharmina, M., Gallego-Schmid, A., Heyes, G. and Azapagic, A. (2017). Integrating Backcasting and Eco-Design for the Circular Economy: The BECE Framework. Journal of Industrial Ecology, 21(3), pp.526-544.

[45] Govindan, K. and Hasanagic, M. (2018). A systematic review on drivers, barriers, and practices towards circular economy: a supply chain perspective. International Journal of Production Research, 56(1-2), pp.278-311.

[46] Despoudi, S. & Dora, M. (2020c) Circular food supply chains. 2020 Institute of Food Science and Technology.

[47] Geng, Y. and Doberstein, B. (2008). Developing the circular economy in China: Challenges and opportunities for achieving 'leapfrog development'. International Journal of Sustainable Development & World Ecology, 15(3), pp.231-239.

[48] Lieder, M. and Rashid, A. (2016). Towards circular economy implementation: a comprehensive review in context of manufacturing

industry. Journal of Cleaner Production, 115, pp.36-51.

[49] Cecc. (2008). Circular Economy Promotion Law of the People's Republic of China (Chinese and English Text) | Congressional-Executive Commission on China. [online] Available at: https://www.cecc.gov/resources/legal-provisions/circular-economy-promotion-law-of-the-peoples-republic-of-china-chinese

[50] Ying, J. and Li-jun, Z. (2012). Study on Green Supply Chain Management Based on Circular Economy. Physics Procedia, 25, pp.1682-1688.

[51] Castellani, V., Sala, S. and Mirabella, N. (2015). Beyond the throwaway society: A life cycle-based assessment of the environmental benefit of reuse. Integrated Environmental Assessment and Management, 11(3), pp.373-382.

[52] Prendeville, S., Sanders, C., Sherry, J., Costa, F., 2014. Circular Economy: Is it Enough?[online] Available at: https://www.researchgate.net/profile/Sharon_Prendeville/publication/301779162_Circular_Economy_Is_it_Enough/links/5727a2be08aef9c00b8b4ddd.pdf

[53] Ying, J. and Li-jun, Z. (2012). Study on Green Supply Chain Management Based on Circular Economy. Physics Procedia, 25, pp.1682-1688.

[54] Henshaw, J., Han, W. and Owens, A. (1996). An Overview of Recycling Issues for Composite Materials. Journal of Thermoplastic Composite Materials, 9(1), pp.4-20.

[55] Storey, J., Emberson, C., Godsell, J. and Harrison, A. (2006). Supply chain management: theory, practice and future challenges. International Journal of Operations & Production Management, 26(7), pp.754-774.

[56] Brandenburg, M. and Rebs, T. (2015). Sustainable supply chain management: a modeling perspective. Annals of Operations Research, 229(1), pp.213-252.

[57] Geng, Y., Zhu, Q., Doberstein, B. and Fujita, T. (2009). Implementing China's circular economy concept at the regional level: A review of progress in Dalian, China. Waste Management, 29(2), pp.996-1002.

[58] Batista, L., Bourlakis, M., Smart, P. and Maull, R. (2018). In search of a circular supply chain archetype – a content-analysis-based literature review. Production Planning & Control, 29(6), pp.438-451.

[59] Mishra, N., Kumar, V. and Chan, F. (2012). A multi-agent architecture for reverse logistics in a green supply chain. International Journal of Production Research, 50(9), pp.2396-2406.

[60] Zorzini, M., Hendry, L., Huq, F. and Stevenson, M. (2015). Socially responsible sourcing: reviewing the literature and its use of theory. International Journal of Operations & Production Management, 35(1), pp.60-109.

[61] Govindan, K. and Soleimani, H. (2017). A review of reverse logistics and closed-loop supply chains: a Journal of Cleaner Production focus. Journal of Cleaner Production, 142, pp.371-384.

[62] Maloni, M. and Brown, M. (2006). Corporate Social Responsibility in the Supply Chain: An Application in the Food Industry. Journal of Business Ethics, 68(1), pp.35-52.

[63] Papaioannou, G, Mohammed, A, Despoudi, S Saridakis, G, Papadopoulos, T. 2020 The role of adverse economic environment and human capital on collaboration within agri-food supply chains, International Journal of Information Management, 52, https://doi.org/10.1016/j.ijinfomgt.2020.102077.

[64] Beske, P., Land, A. and Seuring, S. (2014). Sustainable supply chain management practices and dynamic capabilities in the food industry: A critical analysis of the literature. International Journal of Production Economics, 152, pp.131-143.

[65] Geissdoerfer, M., Savaget, P., Bocken, N. and Hultink, E. (2017). The Circular Economy – A new sustainability paradigm?. Journal of Cleaner Production, 143, pp.757-768.

[66] Despoudi, S. (2019) 8 - Optimized food supply chains to reduce food losses. Saving FoodProduction, Supply Chain, Food Waste and Food Consumption, 227-248.

[67] Despoudi, S, (2021) Challenges in reducing food losses at producers' level: the case of Greek agricultural supply chain producers, Industrial Marketing Management, in press, https://doi.org/10.1016/j.indmarman.2020.09.022